Defence Against Bioterrorism

NATO Science for Peace and Security Series

This Series presents the results of scientific meetings supported under the NATO Programme: Science for Peace and Security (SPS).

The NATO SPS Programme supports meetings in the following Key Priority areas: (1) Defence Against Terrorism; (2) Countering other Threats to Security and (3) NATO, Partner and Mediterranean Dialogue Country Priorities. The types of meetings supported are generally "Advanced Study Institutes" and "Advanced Research Workshops". The NATO SPS Series collects together the results of these meetings. The meetings are co-organized by scientists from NATO countries and scientists from NATO's "Partner" or "Mediterranean Dialogue" countries. The observations and recommendations made at the meetings, as well as the contents of the volumes in the Series, reflect those of participants and contributors only; they should not necessarily be regarded as reflecting NATO views or policy.

Advanced Study Institutes (ASI) are high-level tutorial courses to convey the latest developments in a subject to an advanced-level audience.

Advanced Research Workshops (ARW) are expert meetings where an intense but informal exchange of views at the frontiers of a subject aims at identifying directions for future action.

Following a transformation of the programme in 2006, the Series has been re-named and re-organised. Recent volumes on topics not related to security, which result from meetings supported under the programme earlier, may be found in the NATO Science Series.

The Series is published by IOS Press, Amsterdam, and Springer, Dordrecht, in conjunction with the NATO Emerging Security Challenges Division.

Sub-Series

A.	Chemistry and Biology	Springer
B.	Physics and Biophysics	Springer
C.	Environmental Security	Springer
D.	Information and Communication Security	IOS Press
E.	Human and Societal Dynamics	IOS Press

http://www.nato.int/science
http://www.springer.com
http://www.iospress.nl

Series A: Chemistry and Biology

Defence Against Bioterrorism

Methods for Prevention and Control

edited by

Vladan Radosavljevic
Military Medical Headquarter, Ministry of Defence, Belgrade, Serbia

Ines Banjari
Faculty of Food Technology, Josip Juraj Strossmayer University of Osijek, Osijek, Croatia

and

Goran Belojevic
Institute of Hygiene and Medical Ecology, Faculty of Medicine, University of Belgrade, Serbia

Springer

Published in Cooperation with NATO Emerging Security Challenges Division

Results from the NATO Advanced Research Workshop on Defence against Bioterrorism: Methods for Prevention and Control
Belgrade, Serbia
16–17 March 2017

Library of Congress Control Number: 2018932373

ISBN 978-94-024-1265-9 (PB)
ISBN 978-94-024-1262-8 (HB)
ISBN 978-94-024-1263-5 (e-book)
https://doi.org/10.1007/978-94-024-1263-5

Published by Springer,
P.O. Box 17, 3300 AA Dordrecht, The Netherlands.

www.springer.com

Printed on acid-free paper

Preface

Terrorism is a major threat to the modern world, as for terrorists rules, limits and ethics are nonexistent in the fulfilment of their main goals – horror and panic in a population and social instability and chaos in a state. Bioterrorism is just a matter of choice in this endeavour, the most sophisticated one compared to cold weapons and chemical weapons and firearms. Bioterrorism is "not if, but when and how extensive" it will be. Weaponized biological agents may be available on the black market, and there is also the danger of a potentially dreadful coalition between scientists and terrorists. Since Amerithrax in 2001, there are no proofs of deliberate epidemics, but some unusual epidemiological events such as *Escherichia coli* O104-H4 outbreak in Germany in 2011 were thoroughly investigated on the possibility of a biological attack. On the other hand, pandemics like H1N1 or outbreaks of SARS and H5N1 avian influenza overburdened public health systems in numerous countries and pointed at weak spots of the system. Strengthening and improving counter-epidemic measures at primordial, primary, secondary and tertiary levels is of utmost importance. This is a prerequisite for strong impediment against natural, accidental and deliberate epidemics. Furthermore, intelligence and security are key factors related to timely neutralization of a potential deliberate act.

Political instability, migrations and the rise of extremism are just few of the key factors related to the current menace of terrorism worldwide. Countering bioterrorism is one of the priorities of national health and security systems worldwide. As a contribution to this endeavour, the NATO Science for Peace and Security Programme assigned the editors of this volume the topic "Defense Against Bioterrorism: Methods for Prevention and Control". The top experts worldwide gathered for the NATO Advanced Research Workshop held in Belgrade from 16 to 17 March 2017 and agreed on a creative task of making a multidisciplinary platform against bioterrorism threats and accidental and natural outbreaks. It includes means for differentiation between intentional and natural epidemics, all four levels of prevention, aspects of intelligence and security, preservation of food supply chain, management of panic and ethical aspects.

This scientific volume provides theoretical and practical information on the bioshield against bioterrorism, as well as accidental and natural outbreaks.

Knowledge and professional experience from preventive and clinical medicine, security, intelligence, safety and other areas are systematically classified in primordial, primary, secondary and tertiary levels of prevention. Special emphasis is put on improving the bioshield. Applications of approaches presented in the volume may reduce the possibility of occurrence and consequences of bioterrorism and accidental and natural outbreaks.

The bioshield presented in the volume is highly effective (comprised of four levels of prevention, with additional strategies against bioterrorism), affordable (based on the existing public health and security infrastructure), applicable (may be activated within few hours of an outbreak or a bioterroristic attack) and practical (especially important in decision making). The editors hope that the volume will serve as a tool for development, improvement and/or implementation of the bioshield against bioterrorism.

The editors are thankful to the reviewers for the time and effort invested to provide expert, insightful and constructive suggestions which improved the chapters significantly:

Dr. Gigi Kwik Gronvall, senior associate at the Johns Hopkins Center for Health Security and visiting faculty at the Johns Hopkins Bloomberg School of Public Health
Dr. John McConnell, editor-in-chief of *The Lancet Infectious Diseases*
Dr. Seth Carus, distinguished research fellow at the National Defense University, Center for the Study of Weapons of Mass Destruction, Washington, DC

Special gratitude goes to the NATO Science for Peace and Security Programme for full support in publishing this volume.

Belgrade, Serbia — Vladan Radosavljevic
Osijek, Croatia — Ines Banjari
Belgrade, Serbia — Goran Belojevic
September 2017

Contents

Chapter 1
Introduction

Vladan Radosavljevic

Abstract Bioterrorism is an ongoing threat to global peace and health because weaponized agents are available to potential terrorist groups either directly or through scientists who may be willing to cooperate with terrorists. Biological weapons would have larger disastrous global health effects compared to chemical weapons. Therefore, there is a need for merging endeavors of the scientific community on the tasks of developing methodologies for prevention of biological attack/threat, for biological weapons control and for completing a biodefence system. This monograph is a result of such endeavors.

1.1 Control of Biological Weapons

The scientific community should focus on two strategies:

1.1.1 Strategy of Intelligence

There are no easily identifiable footprints marking bioweapons development as it is nearly indistinguishable from legitimate biological science and biotechnology, and such efforts are easily hidden in plain sight. Gathering intelligence on national level and limiting the illicit transfer of materials, technologies and knowledge, are not easy tasks. Furhter, the intelligence community should develop capability of tactical warning of a planned bioterrorist attack.

In the aftermath of a bioattack the authorities want to know who the attacker was, how the attack was carried out and how the next attack could be stopped? Some of the multibillion dollar biodefence systems, such as Bio Watch in the U.S.A. that has been developed for years does not provide information on who and how attacked,

V. Radosavljevic (✉)
Military Medical Headquarter, Ministry of Defence, Crnotravska 17, Belgrade, Serbia

University of Defence, Belgrade, Serbia
e-mail: vladanr4@gmail.com

V. Radosavljevic et al. (eds.), *Defence Against Bioterrorism*, NATO Science for Peace and Security Series A: Chemistry and Biology,
https://doi.org/10.1007/978-94-024-1263-5_1

will the attack be repeated, and how to stop the next attack. Bio Watch does not provide data in real time (situational awareness) because the system relies on a limited number of sensors that are spread too far from each other. However, situational awareness (information about the nature, location and perpetrator of an attack) is an imperative for informed and rapid decision making [2]. Consequently, BioWatch lacks full efficiency.

So, fighting against biological attacks (BAs) in a classical way, by strengthening intelligence to identify and thwart bioterrorists before they strike, appears to be of secondary importance. The reason is simple: a typical or classical terrorist could only be a perpetrator, while a top scientist may be both, a source of agents and a potential perpetrator, as was probably the case in Amerithrax attack in 2001. A priority should be the development of a cadre of highly skilled and competent analysts to build and maintain biosurveillance systems at all levels. The cheapest and the most effective is an investment in continuing education of personnel on developing new methods and models of biosurveillance. It is necessary to provide a better coordination between the intelligence community and the health research community, and provide more resources to the intelligence community for these efforts.

In this monograph the strategy of intelligence is addressed in the following contributions:

- Synthetic biology, dual use research and proposals for control, authored by Eckard Wimmer, Stony Brook University, NY and University of Göttingen, Germany (USA);
- A Perspective on the Strategy of Intelligence, authored by Randall Murch, Virginia Polytechnic Institute and State University (Virginia Tech), Arlington, VA, (USA);
- The global threats from naturally occurring infectious diseases, authored by Maria – Rita Gismondo, University of Milan, Department of Public Health – Microbiology – Virology Milan (Italy);
- Safety and Security Regulations Against Biological Threats, authored by Anna Bielecka – Oder, The Analysis Team, The Epidemiological Response Center of the Polish Armed Forces, Warsaw (Poland).

1.1.2 Strategy of Deterrence

The first step in fighting against biological attacks are developing pragmatic methods for their detection and identification. At the same time it is the best weapon against hoaxes. It is necessary to develop databases with potential bioagents, as well as prompt and accurate networks for their matching with samples from the focus. Rapid access to accurate and reliable diagnostic data will be of the highest strategic importance in a unusual epidemiological events (UEE). Clinical laboratory data are very specific and reliable, much more than syndromic data or physicians' clinical assessments. Technologies to develop rapid, reliable and cheap diagnostic tests exist (diagnostic tools for public health emergencies), but they should be continually improved.

International efforts should bolster biosurveillance, forensics, training and biosafety, all measures that could lessen the likelihood of biological weapons development and use. Of particular importance are the efforts to strengthen microbial forensics capacity. A need to develop the strongest possible scientific capacity to trace back a pathogen to its natural or laboratory origin is of crucial part of the strategy of deterrence from biological attacks. Although some experts state that the most powerful form of deterrence is the ability to catch a perpetrator and prevent future attacks [2], we do not completely agree with it. Many perpetrators especially "low sophisticated" [3], could be suicidal and catching them is out of question.

In this monograph the strategy of deterrence is the subject of the following contributions:

- The Role of Bioforensics in Medical Bio-Reconnaissance, authored by Lothar Zoller and Gelimer Genzel, Institute for Microbiology of the German Army (Germany);
- Comparison of the Available Methods of Differentiation Between a Biological Attack and Other Epidemics, authored by GoranBelojevic, Institute of Hygiene and Medical Ecology, Faculty of Medicine, University of Belgrade, Belgrade (Serbia);
- Food Safety, Standards and Norms Against Bioterrorism: Food Safety and Hazards, authored by Adela Krivohvalek, Teaching Institute of Public Health "Dr. Andrija Štampar", Zagreb (Croatia);
- Strategic Aspects of Countering Bioterrorism, authored by Katarina Strbac, Ministry of Defence, Belgrade (Serbia);
- The Role of Informal Digital Surveillance Systems Before, During and After Infectious Disease Outbreaks: A Critical Analysis, authored by Avi Magid, Anat Gesser-Edelsburg and Manfred S. Green, School of Public Health, University of Haifa, Haifa (Israel)

1.2 Biothreat and Bioterrorism Prevention

A strategy that anticipates a wide range of possible biothreat&bioterrorism scenarios is necessary. We propose our paradigm with four levels of prevention [3]. Prevention efforts should provide barriers against biological attacks. A well-planned, well-rehearsed and rapidly executed epidemic response can dramatically diminish the consequences of biological attacks. It looks impossible to predict intentions and ways of possible bioterrorist acts. Currently the key matters in developing biodefence system is elaboration levels of prevention – which are unique, systematic and scientific approach. Their aim is to make biodefence more understandable, practical, cheap and easy to use. Approximately 1.8 million airline passengers cross international borders daily, that lead to a free route of radiating infectious biological materials around the world within hours [1]. For the year 2016, World Organization for Animal Health listed 118 animal diseases, infections and infestations that could be developed as bioweapons (www.oie.int). There is no army, police or security agency which may guarantee safety from bioterrorism. So, in the context of biodefence, tasks to every inhabitant are: awareness, surveillance and preparedness.

If considered as a chain to stop biological attack or outbreak, it is necessary to push out as many links of that chain as possible. So, at the beginning we reviewed available methods of differentiation between a biological attack and other epidemics.

1.2.1 Levels of Prevention

Primordial prevention comprises measures and activities to stop entering perpetrator/source of infection /reservoir of pathogen and biological agent/pathogen on defended territory. This is the first line of biodefense, deeply and multiply linked with the strategies of intelligence and deterrence.

The primary prevention of biological attack is focused on monitoring and surveillance of potential internal sources of biological agents and bioterrorists. There are three types of surveillance: clinical (syndromic), laboratory and environmental.

In this monograph the primordial and primary levels of bioterrorism prevention are addressed in the following contributions:

- Primordial and Primary Levels of Biothreat & Bioterrorism Prevention, authored by Vladan Radosavljevic, University of Defense, Ministry of Defense (Serbia);
- Environment and bioterrorism, authored by Stephanie Watier-Grillot et al, French Army, Marseille (France);
- Food and Bioterrorism – The Case of Airline Catering, authored by Ines Banjari, University of Osijek, Faculty of Food Technology, Osijek (Croatia).

The secondary level of prevention is addressed from two aspects: clinical and public heath approaches during biological attack/outbreak:

- Secondary Level of Biothreat & Bioterrorism Prevention, authored by Vladan Radosavljevic, University of Defence, Belgrade (Serbia);
- Preventative Medicine: research and use of medical countermeasures during an outbreak, authored by Inger Damon, Center of Diseases Control, Division of High-Consequence Pathogens and Pathology, Atlanta (USA);
- Rapid and Low-Cost Tools Derived from Plants to Face Emerging/Re-emerging Infectious Diseases and Bioterrorism Agents, authored by Rosella Franconi et al, Department for Sustainability, Italian National Agency for New Technologies, Energy and the Environment (ENEA), 'Casaccia' Research Centre Rome (Italy).

During UEE there are two types of agents, biological agent and information as agent [3]. Appropriate media guiding is highly important in the case of biologal attack/outbreak (motivating population for vaccination, low enforcement, protection of variety of spoofing, hacking and malevolent cyberwar/infowar/psyop activities). In this monograph, the tertiary level of prevention is covered in the following chapters:

- Panic disorder during a bioterroristic attack, authored by Milan Latas, Institute for Psychiatry, University Clinical Center of Serbia, Belgrade (Serbia).

Finally, in this monograph there are some actual topics of general interest in biodefense:

- Refugee crises as a potential threat to public health, authored by Raynichka Mihaylova- Garnizova, Military Medical Academy, Clinic of Infectious Diseases, Sofia (Bulgaria);
- Ethical aspects of bioterrorism and biodefense, authored by Elizabeta Ristanovic, Military Medical Academy, Institute of Microbiology, Belgrade (Serbia).

In the concluding remarks, the editors provide a summary proposals for building biodefence system on local and global levels. This is the final step in this approach, starting from the methods of UEE detection, identification-differentiation, through control to prevention.

References

1. Drexler M (2010) Institute of Medicine (US). In: What you need to know about infectious disease? National Academies Press (US), Washington, DC
2. Gronvall GK, Bouri N, Rambhia KJ, Franco C, Watson M (2009) Meeting report. Prevention of biothreats – a look ahead. Center for Biosecurity of UPMC, October 6, 2009. http://www.centerforhealthsecurity.org/our-work/publications/prevention-of-biothreats-a-look-ahead
3. Radosavljević V, Belojević G (2009) A new model of bioterrorism risk assessment. Biosecur Bioterror 7:443–451

Chapter 2
Synthetic Biology, Dual Use Research, and Possibilities for Control

Eckard Wimmer

Abstract The anthrax attack on the human population in the United States in 2001/2002 may be considered the naissance of modern bioterrorism. This attack, e.g. the planned killing by means of deadly microorganisms (Bacillus anthracis) caused enormous public concern, because, numerous other deadly agents, now known as "select agents", occur in nature and are available for misuse. The anthrax attack coincided with the first report in 2002 of the de novo synthesis in the test tube of a pathogenic human virus, poliovirus, that was equally shocking because it indicated that dangerous infectious agents could be produced in laboratories outside of government control. These events were synchronous with the advent of a new discipline, Synthetic Biology, which was an emerging area of research that can broadly be described "as the design and construction of novel artificial biological pathways, organisms or devices, or the redesign of existing natural biological systems." The synthesis of viruses, or more broadly expressed: each experiment in Synthetic Biology, fits the definition of "Dual Use Research" – the dual use dilemma in which the same technologies can be used for the good of humans and misused for bioterrorism. In view of these threats the US Government has formulated rules that can lower the chances of misuse of biological research. That includes all research with select agents or the modification of agents to acquire dangerous traits ("Gain of Function"). It also calls for the continuous education of all generations entering research: to be aware that results of research can be dangerous, if not immediately then possibly at later times.

E. Wimmer (✉)
Stony Brook University, Stony Brook, NY, USA
e-mail: eckard.wimmer@stonybrook.edu

V. Radosavljevic et al. (eds.), *Defence Against Bioterrorism*, NATO Science for Peace and Security Series A: Chemistry and Biology,
https://doi.org/10.1007/978-94-024-1263-5_2

2.1 Terrorism and Bioterrorism in the USA in 2001

On September 11, 2001, the Unites States suffered the deadliest (non-military) terrorist attack of all times when high jacked airline jets crashed into the World Trade Center Towers in New York City, into The Pentagon in Washington, DC, and into a field in rural Pennsylvania.

About seven days after the 9/11 attack, anonymous letters laced with deadly spores of the *Bacillus anthracis* (Anthrax) arrived in offices of media companies and the US Congress. During the following 2 months at least ten letters, anonymously mailed, were identified causing at least 17 anthrax infections and altogether 5 casualties. This was the first incidence of what seemed to be the planned killing by means of deadly organisms and, as such, may be considered the emergence of modern bioterrorism. The attack triggered a nine-year investigation that consumed many millions of dollars and was extraordinary for its magnitude [1].

The 9/11 terrorist attack claimed >3000 casualties and changed life in the US. The Anthrax attack, in contrast, caused "only" 5 casualties but it was a most frightening wake-up call for the public. Bioterrorism executed with an anonymous, highly dangerous, self-replicating infectious agent, which cannot be detected even with high-powered light microscopes (e.g. viruses), suddenly presented to the society a new vulnerability that could cause deadly chaos. To lower the risk of such scenario all dangerous microorganisms, known now as "select agents", are currently controlled for research by the government and their use is subject to strictest rules.

2.2 Synthetic Biology and Bioterrorism

The publication in July 2002 of the first chemical synthesis of poliovirus [2, 3], an organism harmful for humans, struck a sensitized and anxious American public after 9/11. It became instantaneously clear that viruses like poliovirus no longer "exist" only in nature but also in computers. Viruses, therefore, can be synthesized using the information stored in computers. These synthetic viruses have the computer as parent – a natural isolate is no longer required.

This scenario immediately brings us to an ancient dilemma in science called "*dual use research.*" Every result produced in science has the propensity to be useful and/or harmful for humankind. In addition, every progress in biomedical technology yields the possibility to achieve new "unheard-of" results, like the chemical synthesis of poliovirus. Currently, there are the genomes of >2,500 viruses available in public databases. All of these sequences could be used to synthesize the corresponding viruses, including small pox virus. That is the dark side of progress. However, true to the tenet of "dual use research": the *de novo* synthesis of poliovirus has also lead to the development of very promising new strategies to develop vaccines [4, 5] that will hopefully aid humankind in the future.

Since the first ground breaking synthesis of poliovirus in 2002, some very dangerous viruses have been "re-created" in the test tube in the absence of natural isolates [6]. Most notably was the resurrection of the type A strain of influenza virus 1918/1919 that killed an estimated 20–50 million humans [7]. The virus vanished after the catastrophic pandemic, leaving no trace for investigations that would explain its extraordinary virulence. Studies carried out after its *de novo* synthesis in 2005, however, have identified the murderous armor of this virus and, thus, prepared medical scientists for its possible recurrence. These results justified the enormous efforts to "revive" this terrible virus.

The revival of another lost virus in the absence of the natural template was described in a recent, unpublished report [8]. Briefly, horsepox virus, an orthopoxvirus , thought to be extinct from nature, has been synthesized de novo. This virus, which is related to the human smallpox virus, may have been part of the original vaccine against human smallpox and, therefore, it may be highly useful for improving the current smallpox vaccine. Please note that in the history of humankind smallpox virus is considered the most vicious killer of all time. It was globally eradicated in 1980. Not surprisingly then, the report that the related horsepox virus was rebuild by genetic engineering fueled again the debate about the benefits and risks in biomedical research [8].

In the mid 2010s these events coincided with the advent of a new discipline, Synthetic Biology. Synthetic Biology is an emerging area of research that can broadly be described "as the design and construction of novel artificial biological pathways, organisms or devices, or the redesign of existing natural biological systems" [9]. About 12 years ago, the idea of combining molecular genetics, genetic engineering, cell biology, mathematics and engineering has been heralded as the dawn of a new science with unlimited possibilities to create new artificial biological systems useful for humans and the environment. It has been greeted with skepticism by a nervous general public as these goals could lead also to new bioterrorist agents.

Are there sure measures for preventing misuse by modern biomedical techniques leading to results harmful for humans? An answer provided by Joshua Lederberg (1998 Nobel Prize Laureate in Medicine) is sobering: *"There is no technical solution to the problem of biological weapons. It needs an ethical, human and moral solution – if it is going to happen at all"* [10]. It is worth noting that this crucial statement was made just 3 years before the anthrax attack.

Listed below are some constraints that show how in the US the development of dangerous infectious agents, referred to as "select agents", is controlled – perhaps misuse even prevented – through technical and administrative hurdles:

I. Re-creating an already existing dangerous virus for malicious intent is a complex scientific endeavor. (i) It requires considerable scientific knowledge and experience and, more importantly, considerable financial support. That support usually comes from government and private agencies (NIH, NAF, etc.), organizations that carefully screen at multiple levels all applications for funding of ALL biological research. (ii) It requires an environment suitable for experimenting with dangerous infectious agents (containment facilities). Any work in containment facilities is also carefully regulated.

II. Genetic engineering to synthesize or modify organisms relies on chemical synthesis of DNA. Synthesizing DNA is automated and carried out with sophisticated, expensive instruments. The major problem of DNA synthesis, however, is that the product is not error-free. Any single mistake in the sequence of small DNA segments (30–60 nucleotides) or large segments (>500 nucleotides) can ruin the experiment. Companies have developed strategies to produce and deliver error free, synthetic DNA, which investigators can order electronically from vendors, such as Integrated DNA Technologies (US), GenScript (US) or GeneArt (Germany). This offers a superb and easy way to control experimental procedures carried out in any laboratory: the companies will automatically scan ordered sequences in extensive data banks to monitor relationship to sequences of a select agent. If so, the order will be stalled until sufficient evidence has been provided by the investigator that she/he is carrying out experiments approved by the authorities. The entire complex issue of protecting society from the misuse of select agents has been discussed in two outstanding studies [11, 12].

III. Engineering a virus such that it will be more harmful (more contagious, more pathogenic) is generally difficult because, in principle, viruses have evolved to proliferating maximally in their natural environment. That is, genetic manipulations of a virus often lead to loss of fitness that, in turn, is unwanted in the bioterrorist agent.

A special case, however, is research leading to "Gain-of-Function" of microorganisms, an objective broadly defined by the US Government as follows:

U.S. Government Gain-of-Function Deliberative Process and Research:

Gain-of-function studies, or research that improves the ability of a pathogen to cause disease, help define the fundamental nature of human-pathogen interactions, thereby enabling assessment of the pandemic potential of emerging infectious agents, informing public health and preparedness efforts, and furthering medical countermeasure development. Gain-of-function studies may entail biosafety and biosecurity risks; therefore, the risks and benefits of gain-of- function research must be evaluated, both in the context of recent U.S. biosafety incidents and to keep pace with new technological developments, in order to determine which types of studies should go forward and under what conditions.

US. Government, October 17, 2014.

The debate about Gain-of-Function Research became particularly intense in 2013 when benefits vs risks were broadly assessed in studies of the host range change, increased transmissibility and pathogenicity of non-human infectious agents such as SARS, MERS or animal influenza virus strains. Indeed, following an intense debate, all experiments aimed at "Gain-of-function" of infectious agents were put on hold but have since been re-assessed [13].

2.3 The Responsibility of Scientists to Prevent Bioterrorism

If large, secret organizations (like the "Aum Shinrikyo" in Japan) or governments of hostile countries would plan to develop and apply agents of bioterrorism the opportunities to interfere with such activities would be limited. Therefore, it is imperative to train young scientists to be aware of molecular techniques to modifying existing dangerous organisms or generate an infectious agent with properties of a select agent. This training must also include education in the ethics of science – preparing young scientists to share responsibilities to work only for the good of humankind. Obviously, this is a long-term goal. In the short-term, vigilance and communication are our best defense [14–16].

References

1. Amerithrax or Anthrax Investigation (2010) FBI.gov
2. Cello J, Paul A, Wimmer E (2002) Chemical synthesis of poliovirus cDNA: generation of infectious virus in the absence of natural template. *Science* 297:1016–1018
3. Wimmer E (2006) The test-tube synthesis of poliovirus: the simple synthesis of a virus has far reaching societal implications. EMBO Reports (Special Issue) 7:S3–S9
4. Coleman JR, Papamichail D, Skiena S, Futcher B, Wimmer E, Mueller S (2008) Virus attenuation by genome-scale changes in codon pair bias. Science 320(5884):1784–1787
5. Shen SH, Stauft CB, Gorbatsevych O, Song Y, Ward CB, Yurovsky A, Mueller S, Futcher B,.Wimmer E. (2015). Large-scale recoding of an arbovirus genome to rebalance its insect versus mammalian preference. Proc Natl Acad Sci U S A 112(15):4749–4754. doi: https://doi.org/10.1073/pnas.1502864112. Epub 2015 Mar 30. PMID: 25825721
6. Wimmer E, Song Y, Gorbatsevych O (2015) The future of synthetic virology. In: Katze MG, Korth MJ, Law GL, Nathanson N (eds) Viral pathogenesis: from basics to systems biology. Academic Press, London, pp 320–322
7. Tumpey TM et al (2005) Characterization of the reconstructed 1918 Spanish influenza pandemic virus. Science 310:77–80
8. Kupferschmidt, K. How Canadian researchers reconstituted an extinct poxvirus for $100,000 using mail-order DNA. Science July 2017
9. The fourth international meeting on synthetic biology (SB4.0), Hong Kong 2008
10. Preston R. 1998. Annals of warfare: the bioweaponeers. The New Yorker, March 9, pp 52–65
11. National Research Council (2010) Sequence-based classification of select agents: a brighter line. National Academy Press, Washington, DC. 234 pp
12. Garfinkel M, Endy D, Epstein G, Friedman R (2007) Synthetic genomics: options for governance. Biosecur Bioterror 5:359–362
13. Gain-of-function research: summary of the second symposium, March 10-11, 2016. National Academy of Sciences. see <Gain-of-function Research NASP 2016>
14. Federation of American Scientists (2005) Case studies of dual use research. http://www.fas.org/biosecurity/education/dualuse/FAS_Introduction/1_A.html
15. Presidential Commission for the Study of Bioethical Issues (2010) Creation of the commission. Commission Charter. http://www.bioethics.gov/documents/Charter-of-the-Presidential-Commission-for-the-Study-of-Bioethical-Issues-03.10.10.pdf
16. Proposed framework for the oversight of dual use life sciences research: strategies for minimizing the potential misuse of research information. NATIONAL SCIENCE ADVISORY BOARD FOR BIOSECURITY 2007

Chapter 3
The Global Threats from Naturally Occurring Infectious Diseases

Alessandro Mancon, Davide Mileto, and Maria Rita Gismondo

Abstract Biological risk relates to a broad spectrum of possible scenarios, that can be classified in three categories: natural occurring, unintended and deliberate. The prevention and management of such events require dedicated measures at national and international level, in terms of biosafety and biosecurity: an optimized intervention can minimize the probability of occurrence, but also adverse short-term (i.e.: number of casualties, population reaction…) and long-term (i.e.: chronic illnesses, ecological changes, trades drop…) consequences. Natural scenarios include common, emerging/re-emerging and chronic infectious diseases: they are caused by biological agents, which can be normally present in the communities, as acute or chronic pathologies, or suddenly appear, causing new or uncommon syndromes. In particular, a lot of environmental and human factors can influence emerging and re-emerging diseases: for example, urbanization and people mobility facilitate microorganisms spread, while climate changes are likely to induce a relocation of pathogens vectors. Unintended events are usually due to research and diagnostic activities: laboratories are the places where biological agents are handled and a lack in Biosafety measures or negligence can result in accidental release; the so called Laboratory Acquired Infections represent the main consequence, since they cause pathologies in the laboratory workers, but could be also transmitted in the population. Deliberate use of biological agents is strictly related to terroristic activities: microorganisms are very suitable for this purpose, since they are hidden and can easily spread. The present chapter summarizes the main characteristics of biological

---

Alessandro Mancon and Davide Mileto equally contributed to this work.

A. Mancon • D. Mileto
Laboratory of Clinical Microbiology, Virology and Bioemergencies, ASST Fatebenefratelli-Sacco, "L. Sacco" – University Hospital, Milan, Italy
e-mail: alessandro.mancon@unimi.it; davide.mileto@unimi.it

M. R. Gismondo (✉)
Laboratory of Clinical Microbiology, Virology and Bioemergencies, ASST Fatebenefratelli-Sacco, "L. Sacco" – University Hospital, Milan, Italy

University of Milan, Milan, Italy
e-mail: mariarita.gismondo@unimi.it

V. Radosavljevic et al. (eds.), *Defence Against Bioterrorism*, NATO Science for Peace and Security Series A: Chemistry and Biology,
https://doi.org/10.1007/978-94-024-1263-5_3

agents related events, taking in account their origin and the principal consequences on the community.

3.1 Introduction

Nowadays, when speaking about Biological risk we have to take in account a broad range of possible scenarios, varying from the simple exposure to a microorganism responsible for an infection to an intentional release of biological agents to cause epidemics.

Specifically, the wild spectrum of these risks can be attributable to different events with dissimilar levels of intent: natural occurring, unintended and deliberate (Fig.3.1).

The category of events naturally occurring is represented by all the naturally occurring infectious diseases of humans, animals and plants at national and/or international levels, jointly with the re-emerging ones often responsible of outbreaks that occur cyclically over time.

The unintended events represent a wild category in which it is possible identified different levels of responsibility for the involved staff. Often, the events can be the unintended consequence of a research or in other case can be the consequence of a laboratory accidents in which untrained staff is involved. Within the unintended event the lack of awareness, the underestimation of the risk as well as the negligence by the workers during the conduction of procedures for the management of biological agents represent a big source of risk for the occurrence of unintended event.

The deliberate misuse represents the last zone into the broad range of the biological risk. All the aimed actions to intentional use of pathogens for harmful purposes are include into this last part of the risk spectrum.

As a consequence of the different nature of the biological risk, its management focuses on different sectors:

- preparing for the management of the naturally occurring disease outbreaks, on a national or international scale, facing the impact of the events on the individuals and public health, national and international economies, and social systems;

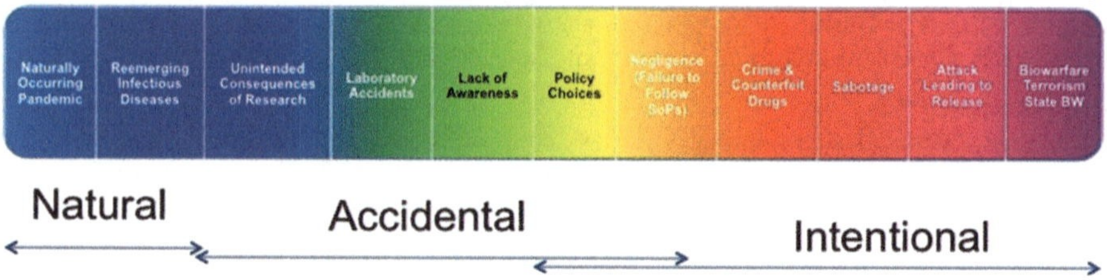

Fig. 3.1 The spectrum of Biological risk (Available on http://iclscharter.org/)

- *Biosafety* defined by the UN as "principles, technologies, practices and measures implemented to prevent the accidental release of, or unintentional exposure to pathogenic agents" [1];
- *Biosecurity*, which refers to the "protection, control and accountability measures implemented to prevent the loss, theft, misuse, diversion or intentional release of pathogenic agents and related resources as well as unauthorized access to, retention or transfer of such material" [1].

3.2 Consequences of Biological Events

The occurrence of a biological event, both on national or international scale, may have social implications with consequence in the short or in the long run.

3.2.1 Short-Term Consequences

Subsequently the occurrence of a biological, local and/or national institutions implement an emergency plan as first response throw specific preparedness strategies with the increasing of medical resources and infrastructures.

Naturally, the different degree of the emergency response depends on the availability of resources, know-how and the specific skills to manage the emergency of each different context in which the emergency occurs.

The consequence of a biological event in terms of number of casualties, also depends on the essential features of the biological agent involved: infectivity, virulence, incubation period, transmission, pathogenicity, lethality and stability, surely are all key features that play a crucial role in defining the severity of the outbreak. This is certainly true especially in the early moments following the emergency breakdown when not all the emergency containment and management measures are already fully active.

Within of a Bioemergency scenario, the negative psychological reactions are unpredictable and the perception of a real biological risk triggers terror and panic within the population.

It is for that reason that the communication is not just the conveyance of information but media should be seen as a partner in supporting risk communication to bridge the gap between the perception of the public and the scientific assessment of risks [2].

3.2.2 *Long-Term Consequences*

The elements affecting the seriousness of the event are several: density of population of the affected area, geographical features of the affected area, quantity of dispersed material and health-care response efficiency [3]. Many of such parameters have a cascade effect which is reflected on the affected community by the event in terms of economic impact, geographical extension, quality of human health, etc., i.e. long-term effects.

According to WHO, effects of biological events may extend beyond their immediate target both in time and space and include: chronic illness, delayed effects, new infectious disease becoming endemic, effects mediated by ecological changes.

Considering a hypothetical scenario in which a biological event could affect plants or animals, the effects could lead to strong ecological changes that could be reason of reduction in the quality and quantity of food supply derived from plants or animals.

3.3 The Spectrum of Biological Events

As mentioned above, the biological risk is linked to a wide range of occurrences, that could be natural, accidental or intentional. Every day population is exposed to common pathogens, that cause many different diseases, such as influenza or gastro-intestinal syndromes, while an unintentional exposure to microorganisms more or less dangerous is usually related to research and health activities; fortunately, the deliberate misuse of biological agents has been infrequent among human history, but the low control on these agents in some countries (i.e.: scarce resources, war…) and the rising number of terroristic activities represent a dangerous scenario also from this point of view.

3.3.1 *Naturally Occurring Events*

Infectious diseases are a stable travel companion for human kind, as demonstrated by the description of such syndromes in ancient times: antique Egyptian paintings depict people with typical poliomyelitis skeletal deformations [4], while during the fifth century BC, in Greece, Thucydides and Hippocrates described different epidemics, the Plague of Athens and the Cough of Perinthus respectively [5]. Many pathogens disappeared, other arise and some other still go with humans, animals and plants.

In order to describe contemporary infectious diseases, we can classify them into three categories: common, emerging/re-emerging and chronic.

Common infections are normally present in the community, in which they induce pathologies of variable severity, according to host and agent characteristics. A main aspect to be considered is the immune system of the affected individual: in normal conditions, it is enough to contrast pathogens action, acting through cell response and antibodies; on the contrary, in immunocompromised patients the infections are likely to evolve to worse conditions and increase to a life-threatening state [6]. In developed countries the use of drugs and public health measures succeeded in contrasting many of these diseases, which remain a big issue in low income nations: the best example are AIDS, malaria and tuberculosis, controlled in Europe, Northern America and other rich areas, while recognized as principal enemies in the other world region by World Health Organization [7].

Emerging and re-emerging infectious diseases are a main global concern, since many of them involve highly hazardous pathogens or a very large population. The US National Institute of Health defined the Emerging infectious diseases as those never occurred in humans before or occurred affecting few individuals in isolated communities or occurred throughout human history but only recent recognition of the infectious etiology. Similarly, Re-emerging infectious diseases are those that were a major health problem, declined dramatically and are now becoming again a concern in a significant population [8].

Swine Flu, Severe Acute Respiratory Syndrome (SARS), Middle East Respiratory Syndrome (MERS) and Ebola Virus Disease (EVD) are perhaps the most well-known examples of emerging diseases, due to the large spread of information through mass media.

Swine Flu emerged during 2009 in Mexico: researchers isolated in humans a H1N1 Influenza A virus, founding a history of contact with pigs in the first cases [9]; the pandemic declared by WHO in the same year resulted in a limited number of infections and deaths: European Centre for Disease Prevention and Control estimated in January 2010 approximately 14,300 deaths Worldwide, very low if we compare with Spanish Flu accounting for more than 100 millions [10].

SARS and MERS are respiratory syndromes, both caused by Coronaviruses. The first was discovered in 2003 in China, where an American business man was hospitalized for an atypical pneumonia: the Italian microbiologist Carlo Urbani, analyzing the clinical case, understood that a new infectious disease was the cause of the illness and permitted to identify a new virus; unfortunately, he contracted the infection, dying after few months [11]. MERS appeared in Middle East area in 2012: a new Coronavirus was isolated from patients in Saudi Arabia, Gulf region and Jordan, causing 34 cases as of May 2013, with a mortality rate of 60% [12]. Such high mortality decreased over time, reaching the actual estimated value of 35%, but WHO experts believe that it is an overestimated data, due to lack in cases reporting; moreover, the restricted geographical region interested suggests a hard human-to-human transmission [13].

Ebola virus belongs to *Filoviridae* family, which also includes *Marburg* virus: these pathogens are known to be highly infectious and to cause severe syndromes characterized by hemorrhages. After their discovery in 1976 and 1967, several epidemics have been reported in Central Africa: these events typically involved a small population, owing to isolation of rural areas in which occurred, but the mortality rates ranged from 50% to 90% [14]. However, in 2014 in the Western Africa Guinea country a new epidemic started, assuming a completely new aspect: the infection spread to border nations Sierra Leone and Liberia, then to Nigeria, Senegal and Mali and it was also imported in US and Europe, with patients repatriated to be cured or, in the US, by a traveler. The epidemic was declared an international health concern and lasted 2 years, like never before, with number of cases and deaths increased up to more than 28,000 and 11,000 respectively [15].

Considering re-emerging diseases, malaria and bacterial infections are considered major concerns. Antimalarials, antibiotics, swamplands reclamation, hygiene practice and other measures reduced the incidence of these pathologies in high-income countries, but also achieved important results in low-income ones. However, some reports underline how the risk of raise of plasmodium and bacteria is increasing: in a book of Organization for Economic Co-operation and Development (OECD) on environmental changes, the authors underlined that the population exposed to the risk of malaria infection will increase up to 5.7 billions in 2050, if no action is take [16]; in the same way, O'Neil and colleagues pointed out that, if the insurgency of antibiotics resistant bacteria is not stopped, in 2050 we will see ten millions deaths ascribable to bacterial-related pathologies, a situation similar to the pre-antibiotics era [17].

Many common diseases have the potential to become an emerging or re-emerging one, certainly because biological agents evolve and adapt to new conditions, but also environment and human action play a main role in these event. Among the multiple factors influencing this process, the followings are the main actors.

- *High mobility of population*: the high number of daily flights, the availability of cars and roads, the presence of efficient railways and the use of touristic boats allow people to rapidly move inside their countries and across them. Such a high mobility helps biological agents, since they travel with people as hidden hosts.
- *Urbanization*: due to urbanization, deforestation causes the loss of natural habitat for many animals, that could be reservoirs of biological agents: since cities take place of forests, the probability of contact between humans and these animals increases, augmenting the risk of microorganisms transmission.
- *Climate change*: many world areas are experiencing a change in temperatures and climates, in particular becoming hotter and more raining. Such events favor the expansion of vectors, like mosquitoes and ticks, in places where they have been eliminated or haven't been ever present: the consequence is a higher probability of infection in populations which were exposed to such risk only in the past or have never been exposed/affected.
- *Encroachment of habitats*: high density of population means more contacts, which in turn entail higher risk of transmission.

3.3.2 Accidental Release of Biological Agents

Accidental releases of biological agents are certainly a relevant issue in Biosafety.

The risks in a health-care facilities are related to the presence of a series of danger factors, both material and procedural, such as: agents (chemical, physical and biological), equipment (high voltage equipment, centrifuges, pressure and vacuum systems, high and low temperatures, needles, sharps), space (crowding or space limitation), organizational-management aspects, lack of information, education and training of the staff (especially the external personnel).

The organization-management aspects are often underestimated: sometimes communication between workers might be difficult and the facilities are often lacking of internal procedures.

As it is often the case, the simultaneous presence of different staff in the same work space can contribute to increasing the risk of an accident: researcher or technical-scientific staff and sometimes external personnel as adjunct researchers, graduate students, fellows or guests are contemporary present in the same room.

The main accidents that may occur in a health-care facilities are: spills of materials, projection of infected liquids from a pressurized device or equipment, break of tubes in a centrifuge, projection of liquid in the eye, cut or prick when handling contaminated materials, bite by a laboratory animal, accidental injection of a contaminated solution, wound or loss of consciousness in a laboratory [18].

As a part of biological accident, surely the Laboratory-Acquired Infections (LAIs) occupy a major role. LAIs also called occupational illness or laboratory-associated infections are defined as all infection acquired through laboratory or laboratory-related activities regardless whether they are symptomatic or asymptomatic in nature [19].

The infections may happen subsequently to an exposure of the biological agent that can be through the respiratory tract, mucous membranes, oral intake, or percutaneous methods.

This is not a new phenomenon for the microbiological laboratories and different type of facilities can be involved: clinical laboratories, research laboratories, production installations.

To determine whether the micro-organism responsible of the worker's disease is present in the laboratory only or if it is present also in the community sometimes might be difficult.

LAIs are also a public health problem because the infected worker may represent a source of infection for his colleagues, relatives, family members or other citizens.

An exhaustive report on LAIs doesn't exist but only voluntary reports of cases study are available. The problem of undeclared laboratory events is widely acknowledged due to fear of reprisal and the stigma associated with such events [20].

During the years, a large number of LAIs were documented and from 2000 the most frequent acquired diseases are: brucellosis, Q fever, hepatitis, tularaemia, tuberculosis and psittacosis [19].

Among the accidental release of biological agents occurred over the years, two cases are certainly noteworthy. The first regards a case of SARS laboratory acquired infection occurred in September 2003 in a BSL-3 laboratory of Singapore when a 27 years old microbiology student got infected due to a laboratory incident: *"From the results of the epidemiologic investigation surrounding the recent case of SARS, it appears that inappropriate laboratory standards and a cross contamination of West Nile virus samples with SARS coronavirus in the laboratory led to the infection of the doctoral student. No evidence could be found of any other source of the infection. West Nile virus and SARS coronavirus were detected in the virus samples handled in the laboratory. There is no evidence of secondary transmission and this is an isolated case of SARS"* [21].

The second case regards an environment accidental release of biological agents due to an error in effluent control of laboratories in Europe regard the release of Foot-and-mouth disease virus.

Foot and mouth disease (FMD) is a highly contagious viral disease which affects all cloven-hoofed animals and it has a negative impact on livestock productivity in countries where the disease is endemic [22]: *"In the European Union (EU), at least three different FMD outbreaks were linked to virus escape from laboratories. The incidents in Tübingen (Germany) and in Maisons-Alfort (France) occurred before 1991, when systematic prophylactic vaccination of cattle against FMD was employed in the majority of countries in continental Europe. The third incident occurred in 2007 in Pirbright (United Kingdom)"* [23].

3.3.3 Intentional Release of Biological Agents

During history, many armies unknowingly used biological agents as weapons. The presence of microorganisms was unknown, but their effects were evident: transmissible diseases were killing people and soldiers could spread them to weaken enemies, by contaminating water supplies and cities by means of infected corps.

A significant example is the siege of the Crimean city Caffa, in 1346. An account by Gabriele de' Mussi reports that Tartars hurled plague cadavers inside the town, trying to conquer it; the author also suggested that it was the origin of European Black Death in the fourteenth century, but this hypothesis has not ever been confirmed [24].

During the nineteenth century, the scientific evolution in the microbiology field allowed to understand the pathogenic mechanisms related to bacteria, virus, fungi and parasites, and to find new cures; however, it also paved the way for research on bioweapons. In particular, after the World War I until the end of Cold War, different countries implemented military programs aiming to develop such new war tools. In Japan, the general Shirō Ishii found and headed the Unit 731, which between 1936 and 1945 conducted a lot of experiments with biological agents on alive subjects. This Unit studied the effect of various agents (i.e.: *Y. pestis, B. anthracis, V. colerae...*), directly infecting the prisoners and exposing Manchurian population to these agents using bombs [25]. Similarly, in 1942 the British army started to test the effect of *B. anthracis* on the Gruinard island: aerosols containing the bacterial spores were spread in the environment, in order to infect several sheep and observe the consequences; the animals acquired infections and died, but the epidemic reached also farms on the mainland [26]. The British government first stopped the project and then quarantined the island until 1990, the year of anthrax-free declaration.

The two previous examples underlined the defects of the Geneva Protocol, the first international effort to limit these kind of highly hazardous programs: the Protocol prohibited the use of biological weapons, but not their possession or development [27]. For this reason, in 1972 the Biological Weapons Convention (BWC) was opened to signature and then it entered into force in 1975: this new document establishes that development, production, acquisition, transfer, stockpiling and use of biological and toxin weapons are strictly forbidden [28].

Since the majority of World countries undersigned the BWC, the main concern related to bioweapons is bioterrorism. Terroristic groups aim to generate panic inside the communities and the possible arise of dangerous highly infectious diseases is one of the most frightening nightmare; fortunately, only few attempts occurred, with limited consequences. In 1984, a group of fanatic followers of the spiritual leader Osho Rajneesh used *Salmonella typhimurium* to contaminate food in some restaurants at The Dalles, Oregon: a total of 751 cases of gastroenteritis occurred, among restaurants employees and customers, but no patient died; 2 years after, the suspects were sentenced to prison [29]. A similar case took place in Japan, by the religious sect Aum Shinrikyo. This group is well-known for the Tokyo tube sarin attack in 1995, but they also implemented a bioweapon program, to get various biological agents; in particular, it succeeded in retrieving *C. botulinum* and *B. anthracis* and used them in some unsuccessful attacks: in 1990, the sect members, thinking to spread botulinum toxin, pumped an aerosol in Tokyo, Narita airport and Yokohama and Yokosuka US naval bases, but did not achieve te expected results; 3 years later, they tried twice to spray anthrax from the roof of a building in Tokyo, with the same null result [30]. Perhaps the most known case of bioterrorism is the anthrax one in 2001 in USA: letters containing *B. anthracis* spores were mailed to news media and two senators offices, causing 5 deaths and 17 infections; in 2007, FBI accused the scientist Bruce E. Ivins to be the author of the attack, declaring in 2008 that he was the only culprit, but many doubts exist about the investigations; Ivins committed suicide few days before such declaration.

The recent rising number of terroristic acts underlines the importance of implementing a biosecurity program, at both national and international levels.

3.4 Conclusion

Within the control of infectious diseases, indistinctly caused by natural, accidental or intentional events, the rapid detection of the biological agent and the proper management of the event itself, represent a crucial key point, since emerging, re-emerging, and novel infectious diseases are involved. In an increasingly globalized world, an active monitoring of outbreaks, as well as strategies and resources for an adequate and effective response are required to avoid the spread of such diseases. The 2013–2016 West Africa Ebola epidemic represents an example of how the lack of resources and preparedness can affect the response. The outbreak began in December, 2013, in a Guinean rural area, but the country did not have the capacity to detect it; in March, 2014, several organizations, such as Médecins Sans Frontières (MSF), responded and at the same time new cases were confirmed in the capital Conakry and in the bordering Liberia. During the two following months, the operators managing the epidemic realized that adequate diagnostic tools, protective gears, supportive cures and disease knowledge were missing, increasing healthcare workers and communities exposure; MSF warned about the severity of the problem, but both Guinea government and WHO downplayed it to avoid panic. The consequences became soon evident: the epidemic turned into an out of control scenario, resulting at the end in more then 28,000 infections and more than 11,000 deaths [31]. As positive outcome, WHO quickly reacted to the new Ebola outbreak in the Democratic Republic of Congo in May, 2017: experts, resources and training were provided for emergency management, minimizing the infection spread and increasing the preparedness [32].

The management of emergencies is usually considered only a public health issue, but it very often regards also plants and animals.

Above mentioned events highlight the different effects arising not only from lack in implementation of biosafety and biosecurity measures and adequate funding, but also the insufficient training of the health-care workers and risk awareness; moreover, the descripted cases emphasize the need for a national legislative basis for standardization in laboratories.

Certainly, the increase of internal protocols for laboratory practice, as well as the proper day-to-day maintenance of all high containment equipment and structures could help to contain any possible microorganism spreads. In the same way, continuous training for health-care workers plays a crucial role, in order to reduce errors during the management of the biological agents and to improve the quality of response in case of emergencies.

Also our institution is engaged in these purposes. Thanks to the availability of high level containment facilities, as national referral center, "L. Sacco" University Hospital is directly involved in the management of bioemergencies: the Emergency Room has an isolation triage room for suspect cases management; an high containment isolation unit is present at the Infectious Diseases Department for the hospitalization of confirmed patients; the Clinical Laboratory of Microbiology includes Biosafety Level 3 and 4 laboratories, for the diagnosis of highly hazardous agents.

Moreover, as centre of excellence in the field of Biosafety and Biosecurity, we periodically organize training courses on biosafety, aimed at strengthening the skills of health-care workers, and implement several international projects for biosafety culture spreading.

In conclusion, infectious diseases still represent a main concern for public health and global community; nevertheless, responsible national and international joint efforts are likely to properly face the possible coming emergencies.

References

1. Meeting of the states parties to the convention on the prohibition of the development, production and stockpiling of Bacteriological (Biological) and toxin weapons and on their destruction. Report of the meeting of states parties. Geneva; December 1–5, 2008. BWC/MSP/2008/5. http://www.opbw.org/new_process/msp2008/BWC_MSP_2008_5_E.pdf. Accessed 27 Oct, 2016
2. Dickmann P, Bhatiasevi A, Chaib F, Baggio O, Banluta C, Hollenweger L, Maarouf A (2016) Biological Risks to Public Health: Lessons from an International Conference to Inform the Development of National Risk Communication Strategies Report of an international conference on risk communication strategies before, during, and after public health emergencies, Rabat, Morocco, October 22–23, 2015. Health Security 14(6.) Mary Ann Liebert, Inc. https://doi.org/10.1089/hs.2016.0050
3. Cavallini S, Bisogni F, Mastroianni M (2014) Economic impact profiling of CBRN events: focusing on biological incidents. Arch Immunol Ther Exp 62:437–444. https://doi.org/10.1007/s00005-014-0306-x
4. Galassi FM et al (2017) Poliomyelitis in ancient Egypt? Neurol Sci 38(2):375
5. Martin PMV, Mertin-Granel E (2006) 2,500-year evolution of the term epidemic. Emerg Infect Dis 12(6):976–980
6. Barton FH et al (2014) Introduction to the immune system. In: Kasper D et al (eds) Harrison's principles of internal medicine, cap. 360, 19 edn (electronic version). McGraw-Hill, New York
7. World Health Organization – WHO (2015) Accelerating progress on HIV, tuberculosis, malaria, hepatitis and neglected tropical diseases (HTM). WHO, Geneva. Available from: http://apps.who.int/iris/bitstream/10665/204419/1/9789241510134_eng.pdf?ua=1
8. National Institutes of Health (US), Biological Sciences Curriculum Study; Understanding Emerging and Re-emerging Infectious Diseases; NIH Curriculum Supplement Series [Internet], Bethesda (MD): National Institutes of Health (US) (2007) Available from: https://www.ncbi.nlm.nih.gov/books/NBK20370/
9. Dawood FS, Jain S, Finelli L, Shaw MW, Lindstrom S, Garten RJ, Gubareva LV, Xu X, Bridges CB, Uyeki TM (2009) Emergence of a novel swine-origin influenza A (H1N1) virus in humans. N Engl J Med 361. https://doi.org/10.1056/NEJMoa0903810
10. European Centre for Disease Prevention and Control; ECDC Daily Update – Pandemic (H1N1) 2009–18 January 2010 (PDF)
11. Ksiazek TG et al (2003) A novel coronavirus associated with severe acute respiratory syndrome. NEJM 348:1953–1966
12. De Groot RJ et al (2013) Middle East respiratory syndrome coronavirus (MERS-CoV): announcement of the coronavirus study group. J Virol 87(14):7790–7792
13. World Health Organization – WHO (2017) Middle East respiratory syndrome coronavirus (MERS-CoV) – Fact sheet (updated May 2017); World Health Organization, Media centre, Available from: http://www.who.int/mediacentre/factsheets/mers-cov/en/. Last access 31 May, 2017

14. Centers for Disease Control and prevention – CDC; Ebola (Ebola Virus Disease) – Outbreaks Chronology; Centers for Disease Control and prevention, Available from: https://www.cdc.gov/vhf/ebola/outbreaks/history/chronology.html, Last access 31 May, 2017
15. Centers for Disease Control and prevention – CDC; Marburg hemorrhagic fever (Marburg HF) – Outbreaks Chronology 1967–2014; Centers for Disease Control and prevention, Available from: https://www.cdc.gov/vhf/marburg/outbreaks/chronology.html, Last access 31 May, 2017
16. Organisation for Economic Co-operation and Development – OECD (2012) OECD Environmental Outlook to 2050. OECD publishing. https://doi.org/10.1787/9789264122246-en
17. O'Neill J (2014) Review on antimicrobial resistance antimicrobial resistance: tackling a crisis for the health and wealth of nations. London: Review on Antimicrobial Resistance. Available from: https://amr-review.org/sites/default/files/AMR%20Review%20Paper%20-%20Tackling%20a%20crisis%20for%20the%20health%20and%20wealth%20of%20nations_1.pdf
18. World Health Organization (2004) Laboratory biosafety manual, 3rd edn, Avbailable from: http://www.who.int/csr/resources/publications/biosafety/WHO_CDS_CSR_LYO_2004_11/en/
19. http://www.biosafety.be/HomePage.html
20. Sewell DL (1995 Jul) Laboratory-associated infections and biosafety. Clin Microbiol Rev 8(3):389–405
21. Review Panel on New SARS Case and Biosafety. Biosafety and SARS Incident in Singapore September 2003. Available from: http://www.biosafety.be/CU/PDF/Report_SARS_Singapore.pdf
22. Diego Viali dos Santos et al; Identification of foot and mouth disease risk areas using a multi-criteria analysis approach. PLOS ONE | https://doi.org/10.1371/journal.pone.0178464 May 26, 2017
23. OVERVIEW REPORT Bio-risk management in laboratories handling live FMD virus. EUROPEAN COMMISSION HEALTH AND CONSUMERS DIRECTORATE-GENERAL. Ref. Ares (2015)854105 – 27/02/2015. Available from: http://ec.europa.eu/food/fvo/overview_reports/act_getPDF.cfm?PDF_ID=335
24. Wheelis M (2002) Biological Warfare at the 1346 Siege of Caffa. Emerg Infect Dis 8(9):971–975
25. Sheldon H (2002) Harris; factories of death: Japanese biological warfare, 1932–45 and the American cover-up, 2nd edn. London, Routledge
26. Manchee RJ et al (1981) Bacillua anthracis on Gruinard Island. Nature 294:254–255
27. The United Nations General Assembly; Protocol for the Prohibition of the Use of Asphyxiating, Poisonous or Other Gases, and of Bacteriological Methods of Warfare; Geneva, 17 June 1925
28. The United Nations Office at Geneva; The biological weapons convention. Available from: https://www.unog.ch/80256EE600585943/(httpPages)/04FBBDD6315AC720C1257180004B1B2F?OpenDocument. Last Access 18 Aug 2017
29. Török TJ et al (1997) A large community outbreak of salmonellosis caused by intentional contamination of restaurant salad bars. JAMA 278(5):389–395
30. Smithson AE (2000) Chapter 3: Rethinking the Lessons of Tokyo. In: Smithson AE, Levy L-A (eds), Ataxia: the chemical and biological terrorism threat and the US response. Washinghton, Henry L. Stimson Center, pp 77–79
31. Moon S et al (2015) Will Ebola change the game? Ten essential reforms before the next pandemic. The report of the Harvard-LSHTM Independent Panel on the Global Response to Ebola; The Lancet 385(10009):2204–2221
32. World Health Organization; Emergencies – In equatorial Congo, WHO and its partner respond to an Ebola outbreak; http://www.who.int/emergencies/ebola-DRC-2017/articles/working-with-partners/en/

Chapter 4
Refugee Crisis As a Potential Threat to Public Health

Raynichka Mihaylova-Garnizova and Vasil Garnizov

Abstract The refugee crisis in Europe continues to persist despite recent data, showing a drop in the number of refugees seeking asylum. The EU has called this as "an unprecedented displacement crisis" and has aimed at devising a comprehensive approach to tackle it, which has been widely criticized. Concerns about public healthcare aspects of the crisis have permanently entered the media and policy discourse even though no systematic association between migration and the importation of infectious diseases has been recorded. In this context, the literature has not filled the existing gap between discourse and evidence, and almost no publications with reliable empirical data exist, both thematic (epidemiology) and geographical (Eastern Europe and Bulgaria). Among the existing publications, the focus has been on TB and HIV (Odone et al., Euro J Public Health 25(3):506–512, 2015). In light of this, the aim of this research is to contribute to the debate by providing an overview of the refugee situation in Bulgaria, as a primary entry-point for refugees entering the EU. In order to achieve this, the article analyses the case of the refugee camp in city of Harmanly, close to the Bulgarian-Turkish border, and assesses the public health risks related to this specific situation. Based on a study of 128 patients with different symptoms we aim to draw wider implications about the linkages between public health and migration. The in-depth review of this specific case shows that both the probability and impact of migration on public health increases when the hosting country is relatively poor, the domestic public healthcare system is not efficient, and there is lack of trust in the government and public services. The study contributes to understanding better these risks in order to identify potential mitigation strategies in the region and the EU as a whole.

R. Mihaylova-Garnizova (✉)
Department of Infectious Diseases, Military Medical Academy, Sofia, Bulgaria

V. Garnizov
Department of Anthropology, New Bulgarian University, Sofia, Bulgaria

V. Radosavljevic et al. (eds.), *Defence Against Bioterrorism*, NATO Science for Peace and Security Series A: Chemistry and Biology,
https://doi.org/10.1007/978-94-024-1263-5_4

4.1 Introduction

The European migrant/refugee crisis in 2015–2016 deeply challenged the political, economic and healthcare systems in the European Union (EU). Looking at these developments, we are faced with the key problem of lack of data on the health profiles of migrants, in particular – of refugees. Immediately after the crisis began there were many publications that insist, without extensive evidence, that migrants do not pose a health threat to the host population. Almost no publications could be found with reliable empirical data, especially epidemiological, including for Bulgaria.

The absence of reliable and robust data is even more problematic at times of crisis since public health systems cannot prepare adequately. The lack of objective information also opens up opportunities for the emergence of public myths and psychoses, including disinformation that can lead to political destabilization, which could in turn affect national security.

The purpose of our article is to collect, summarize and present epidemiological data related to migrants in Bulgaria and, on the basis of this information, to analyze the potential risks to public health (including risks to migrants) and to assess the capacity of Bulgaria's healthcare system to cope with the refugee crisis.

4.2 Background

Migration is a phenomenon known since antiquity. Thanks to it, new countries have emerged and new communities have formed. But what is happening today is extraordinary because millions of people constantly change their place of residence for various economic and non-economic reasons, going beyond the scope of our research. The bottom line is that at the end of 2014, almost 60 million individuals were forcibly displaced worldwide. Twenty million of them are refugees. For the first time, Turkey has become the country with the largest number of refugees (Fig. 4.1).

As far as Europe is concerned, with the exception of certain high-income member states, most countries did not have serious difficulties with migrants. This was fundamentally altered after the series of events in North Africa and the Middle East, known as the Arab Spring, as well as the civil war in Libya, and especially since the Syrian conflict began in 2011. Based on regular EASO annual reports, since 2011 one can see the changing structure of migration trends altogether. Until then migration was mainly from lower income countries to countries with a higher income and the greatest percentage of migrants represented the share of those arriving from Western Balkan countries. After 2012, migration started affecting not only high-income countries but also all European countries with the highest percentage of asylum seekers being refugees from Syria (Fig. 4.2).

As a result, Europe has faced a unique influx of refugees, asylum seekers and other migrants since the establishment of the EU: 1.5 million people arrived in the

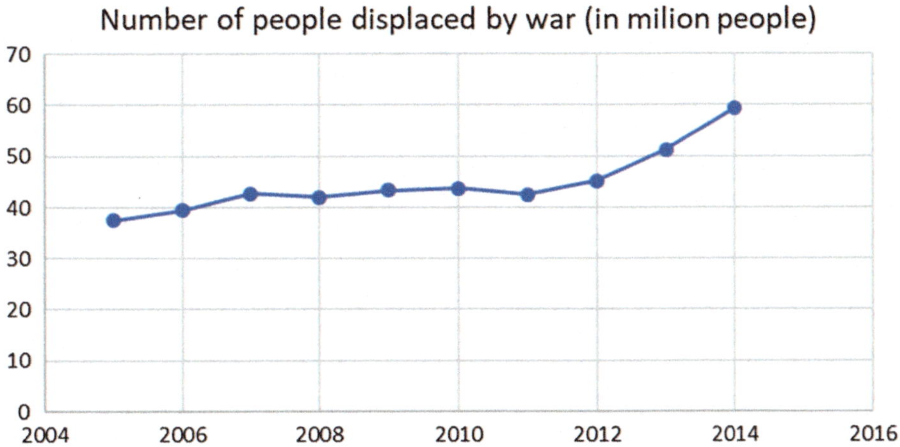

Fig. 4.1 Global migrant growth for 10 years (Source: Adapted following UNHCR [26])

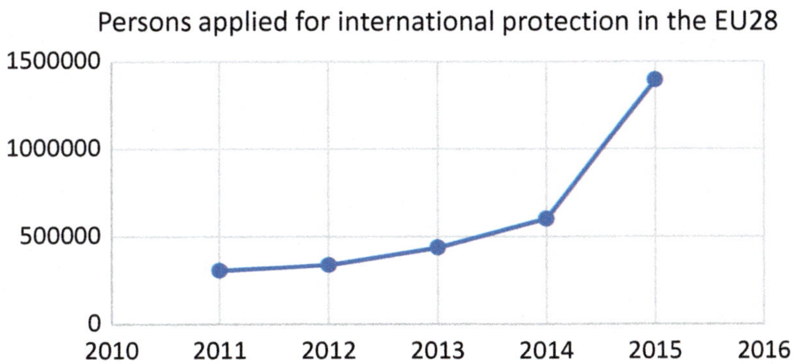

Fig. 4.2 Persons who applied for international protection in the EU28: 2011–2015 (Source: Based on EASO [6–9])

EU only in 2015. Since the beginning of 2016, 1,171,138 applications have been recorded in the EU+ (EU 28 plus Norway and Switzerland). Even though this is 6% less than in the same period of 2015, when 1,242,572 applications were lodged, this highlights the extent of the challenge [10].

This is a not an easy situation to be rationally understood since it is sensitive to strong public opinion and intense debate. Even more, from the point of view of public health 'this refugee situation is unparalleled since the end of WW2' [17]. In this context, often the public healthcare aspects of migrant crises enter the broader public discourse, regardless of expert opinion.

The diverse discourses have been addressed by multiple international organisations. In 2015 Dr. Zsuzsanna Jakab, WHO Regional Director for Europe, clearly highlighted that there is no systematic association between migration and infectious diseases stating that: 'Communicable diseases are primarily associated with poverty. Refugees and migrants are exposed mainly to the infectious diseases that are common in Europe, independently of migration. The risk that exotic infectious agents, such as Ebola virus or Middle East respiratory coronavirus (MERS-CoV), will be imported into Europe is extremely low. Experience has shown that, when it occurs, it affects regular travellers, tourists or health care workers rather than refugees or migrants' [13].

In 2016 the European Commission also addressed the healthcare aspects of migrant crises, declaring that refugees are actually the ones at risk, rather than a burden on health systems. DG SANTE made it explicit that measures to protect refugees' health are being taken 'not out of unfounded fears that they might spread infectious diseases' or 'place a burden on health systems. Their [refugees'] health is at risk, **not the health of EU citizens**' ([22], emphasis added).

WHO and EC's stance on public health challenges of large-scale migration and preparedness of countries in the European region could be seen as too optimistic. WHO insisted that 'the health systems in the countries receiving migrants are well equipped and experienced to diagnose and treat common infectious diseases; they should also be prepared to provide such health care to refugees and migrants. Should a rare exotic infectious agent be imported, Europe is well prepared to respond, as shown over the past 10 years in responses to imported cases of Lassa fever, Ebola virus disease, Marburg virus disease and MERS' [28]. At the same time, events in Germany on the weekend of 12–13 September 2015 demonstrated the opposite: large-scale migration provoked serious malfunctioning of both political and administrative systems including police, social services, housing, public healthcare [1]. After these events, reported in depth by Robin Alexander, the responses across Europe quickly escalated. Once the EU-Turkey deal was agreed, Balkan border closures organised by Austria followed by Hungary, Serbia and Macedonia left some 11–13,000 migrants blocked in the informal camp in Idomeni on the Greek-Macedonian border [24]. From the beginning of March to the end of May asylum seekers there survived in poor sanitary conditions with very limited running water, no lavatories, and insufficient food, despite efforts by NGOs such as 'Hot food Idomeni' and medical care from 'Médecins sans frontière'. The unhealthy and overcrowded conditions at the camp have given rise to infections and in March some cases of Hepatitis A have been reported [24]. Posters with the message 'Greece will offer you accommodation, food and healthcare' (written in Arabic, Farsi and Pashtun) were posted in March but the final decision was taken only on 24th of May. In reality, initial reactions and follow up to the Idomeni situation were limited since there was no confirmed information on the epidemiological situation in Idomeni, no risk assessment and no further public health measures were taken.

Fig. 4.3 Map of Bulgaria

4.3 Bulgaria and Migrant Crisis

In contrast to WHO and EC statements, Bulgarian authorities never declared their readiness to manage a large-scale migrant influx. Until 2010 Bulgaria had not been a country of interest to migrants and, as a result, there was scarce experience of responding to a migrant crisis. Rather, the country was a source of migrants towards Western Europe. Bulgarian membership of the EU on January 1, 2007 led to an increase in the flow of migrants to Bulgaria and a slow but sustainable decrease of Bulgarian emigrants. This situation has changed radically since the start of the civil war in Syria in 2011. The geographical position of the country, which shares a 259-km border with Turkey, makes it a natural primary point of entry for refugees entering the EU (Fig. 4.3). Moreover, Turkey is the largest host country to Syrian refugees.

4.3.1 Demographic Profile of Refugees in Bulgaria

The dramatic increase of refugees in Bulgaria led to a peak in 2013. Compared to 2010, when the number of people legally seeking asylum in Bulgaria was 855, 2013 saw an increase to 7415, a 416% increase from 2012, with the largest share of the migrants, arriving from Syria – 63%. Indeed, migrants account for only 1.6% of migrants in the EU28, but the ratio is approximately 1000 migrants per 1 million people (Fig. 4.4).

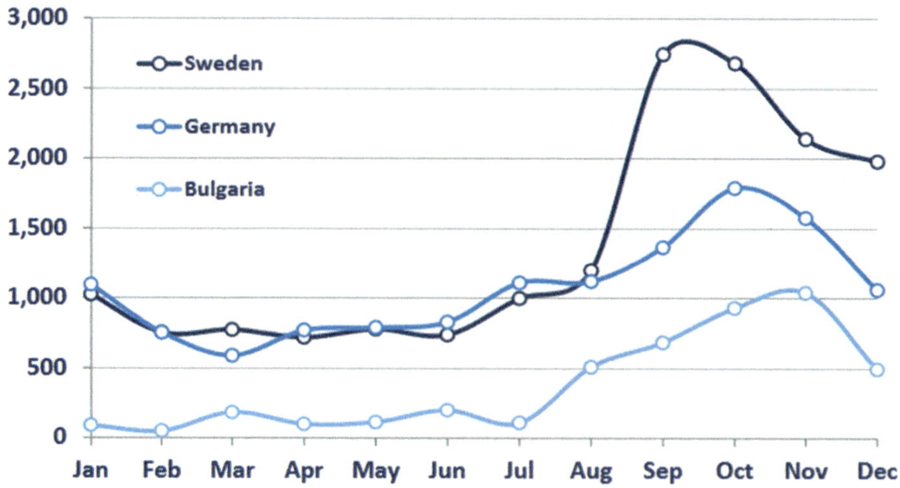

Fig. 4.4 Evolution of Syrian asylum applicants in selected EU Member States, 2013 (Source: EASO [8])

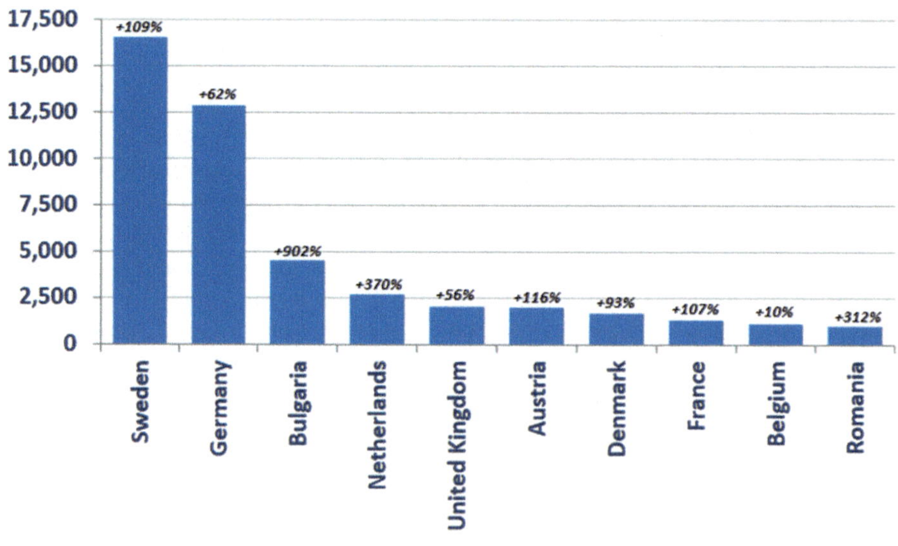

Fig. 4.5 Syrian applicants in 2013 and year-to-year change by main receiving Member States (Source: EASO [8])

For the first time in the recent history of Bulgaria there is a large increase in the flow of migrants. Compared with other Member States, Bulgaria shows the highest rate of growth in the number of Syrian refugees, equal to 902% (Fig. 4.5).

At the same time, the majority of migrants did not intend to stay long in Bulgaria, which they saw only as a stage of their journey to Western Europe. Many of the migrants entering Bulgaria in 2013 moved to France, Germany and Sweden,

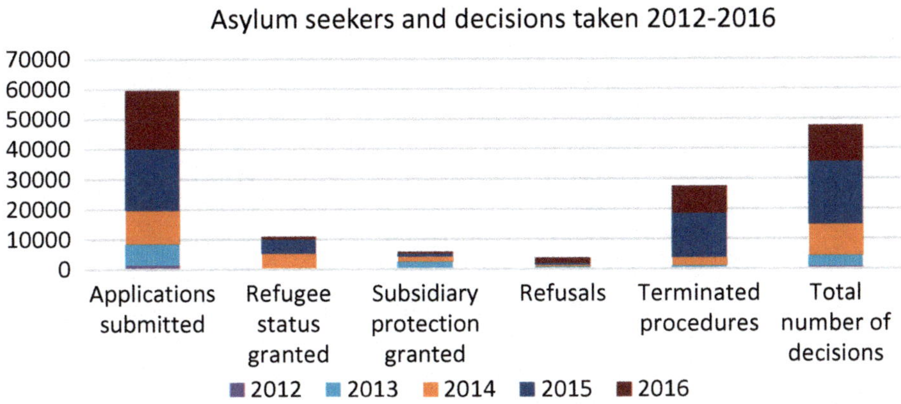

Fig. 4.6 Asylum seekers and decisions taken 2012–2016 (Source: SAR [19])

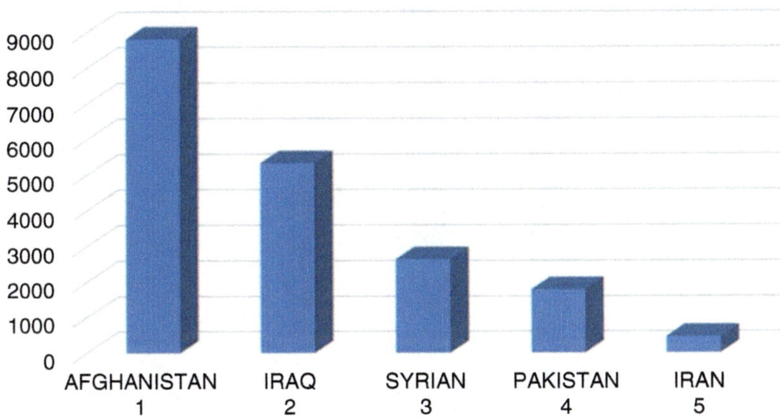

Fig. 4.7 Top 5 asylum seeker countries of origin 2016 (Source: SAR [19])

probably due to the lower incomes, but also the lack of previous settlers and migrant communities [25]. The total number of those remaining in Bulgaria is relatively low, compared to Greece, Italy, and Germany (Fig. 4.6).

It should be noted that for the period from 2012 to 2015, a major share of refugees originated from Syria, while in 2016, the ratio changed, and most of those arriving, were fleeing Afghanistan and Iraq (Fig. 4.7).

In terms of the demographic profile of those arriving, migrants in Bulgaria are relatively less educated and the share of men is approximately equal to that of women and children Figs. 4.8 and 4.9.

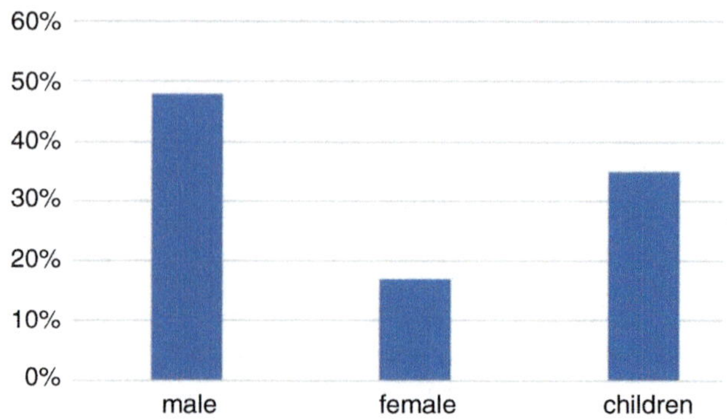

Fig. 4.8 Refugees by gender (Source: SAR [19])

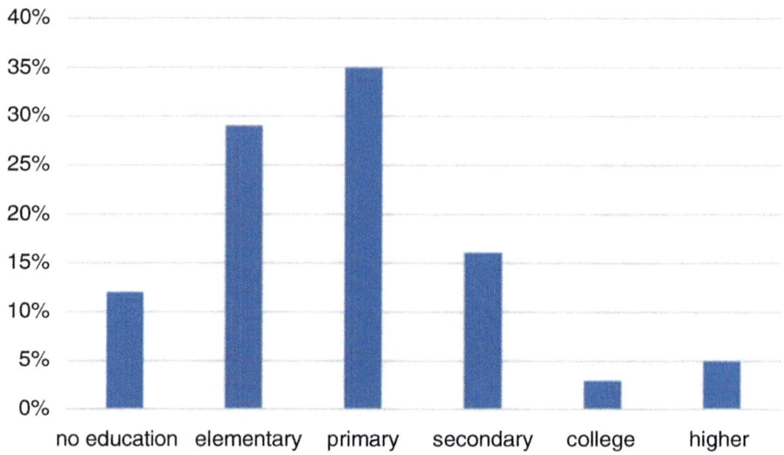

Fig. 4.9 Refugees by education (Source: SAR [19])

4.3.2 Health Aspects of Migrant Crisis in Bulgaria

Following the 2012–2013 crisis, in February 2015 a WHO assessment mission to Bulgaria took place to access the country's capacity to address the public health implications of sudden large-scale influxes of migrants [27]. The mission concluded that the level of assistance provided to the migrants by the Bulgarian Government has clearly improved but the medical response system is very fragile and not fully prepared to respond to a possible new and larger influx of migrants. The main weakness of the response system is the weak provision of primary health care through clinics in migrant centres or the assignment of general practitioners (GP) to migrants due to the fact that clinics are understaffed and interpretation services are not always available. As a consequence, the first recommendation of WHO is the revision of the National plan for crisis management with a focus on reorganizing primary health care services and rationalizing the use of available resources [27, p. 15].

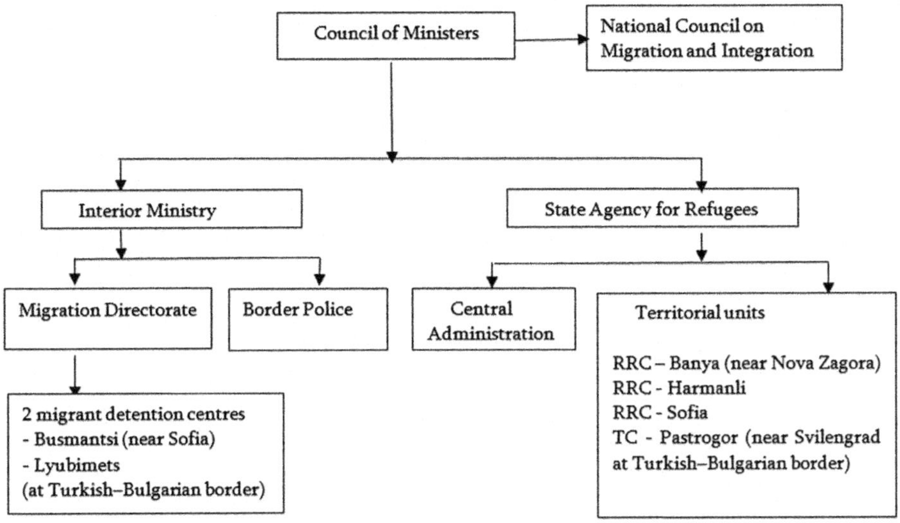

Fig. 4.10 Refugee managing bodies

To put this into context, in the following section we provide an overview of the institutions and personnel involved in refugee and migrant management. Upon arrival in Bulgaria, the asylum seekers first face the authorities of the Ministry of Interior. By law, migrants can be held for up to 24 hours at a location, where basic health screening is conducted, and from there they are directed to the Registration and Reception Centre (RRC), the Transit Centre (TC) or detention centres where primary care is carried out. Primary healthcare in detention centres is the responsibility of the Medical Institute of the Ministry of Interior, situated in Sofia, while medical staff in RRC and TC should be provided by the State Agency for Refugees (Fig. 4.10).

According to Bulgarian legislation, access to public health services requires payment of health insurance contributions. Asylum seekers and persons applying for refugee status are funded by the State Agency for Refugees. In case they have acquired refugee status or the right to asylum, they are obliged to pay health insurance themselves to access the basic package of health care to which Bulgarian citizens are entitled under the insurance system. However, the vast majority of refugees cannot cover their health insurance. For this reason, in March 2017 the Bulgarian government passed a normative act creating an obligation for municipalities to cover health insurance expenses of migrants with official refugee status already granted. After strong public outcry the government withdrew the proposed act and this issue remains open. Moreover, even when migrants fulfil their financial contribution, the number of refugees with an assigned GP is limited due to language barriers and the reluctance of many GPs to take on patients whose residence is likely to be temporary [27, p. 6].

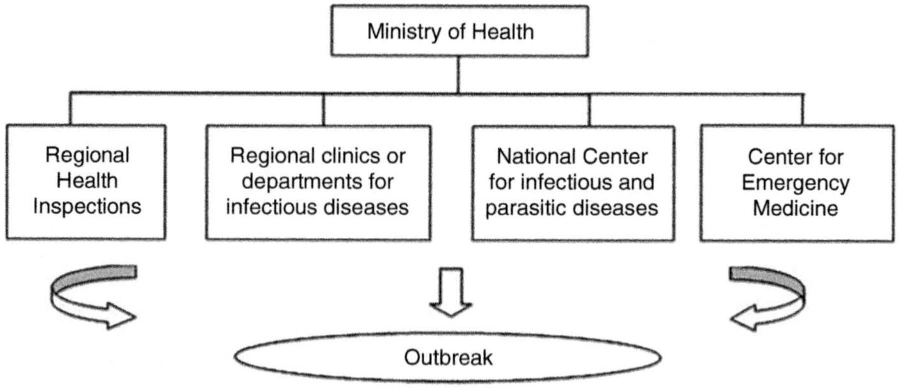

Fig. 4.11 CD surveillance in Bulgaria (Source: Mihaylova-Garnizova and Plochev [23])

The other important recommendation of the WHO assessment mission to Bulgaria is the revision of the disease surveillance early warning and response system, introducing syndromic surveillance to increase early detection of outbreaks and effectively monitoring selected disease trends [27, p. 6]. Such a syndromic surveillance system has not been created yet and it is unknown whether the government plans to implement one. At the moment Bulgaria only conducts surveillance and control of communicable diseases (CD) under the authority of the Ministry of Health (Fig. 4.11).

However, this information system does not include relevant information on the spread of infectious diseases among migrants/refugees. In the current paper we manage to fill the gap with operational reports provided by the Chief State Health Inspector, Angel Kunchev from Regional Health Inspections (RHI). These reports include epidemiological data from *RHI-Haskovo* and *RHI-Sofia* for 2016. Actually, these two administrations cover the two territories of the country in which almost all migrants are situated and thus the data can be seen as representative for the country.

During 2016, refugee centres Harmanli, Pastrogor and Lyubimets, situated on the territory under control of *RHI-Haskovo*, reported 14,901 tests:

- 1877 migrants – 5631 microbiological tests (salmonellosis, dysentery and E. coli);
- 1877 migrants – 3754 parasitological tests (average 2 tests per person - intestinal helminths and intestinal protozoa);
- 728 migrants – 1456 tests for malaria (2 tests per person);
- 87 migrants – 87 malaria with rapid test (1 test per person);
- Unknown number of migrants – 1714 tests for syphilis;
- Unknown number of migrants – 2259 tests for HIV.

Table 4.1 Epidemiological data for 2016 – *RHI-Haskovo*

Diagnosis	# of cases
Scarlet fever	2
Varicella	20
Mumps	1
Enterocolitis	16
Salmonellosis	1
Rotaviral gastroenteritis	4
Acute viral hepatitis B	1
Malaria	9
Echinococcosis of the lung	1

Source: Created on the basis of the report of *RHI-Haskovo*

Based on these tests on the territory of *RHI-Haskovo* in the figure below we illustrate the communicable diseases, which have been diagnosed in 2016 (Table 4.1).

During the same period, *RHI-Sofia*, responsible for the epidemic control of all RRC and detention centres on the territory of the capital, carried out 19,859 tests, as follows (the report does not indicate the number of people examined):

- Parasitological tests – total 13,781 including,
  - Intestinal helminths and intestinal protozoans – 7124;
  - Malaria – 4338;
  - Malaria with an express test – 150;
  - Microfilaria – 2169;
- Microbiological tests – 5676, including:
  - NAG-virions in waste water – 12 samples with negative results;
- Serological tests – total 402, including:
  - Syphilis – 148, with four positive results;
  - HIV / AIDS – 155, with one positive result;
  - HAV Ig M – 3;
  - HAV total – 1 positive result;
  - HbsAg – 46, 8 positives (hepatitis B infection);
  - HCV – 49, 5 positives (hepatitis C infection);

Based on these tests on the territory of *RHI-Sofia* the following communicable diseases have been diagnosed (Table 4.2).

Despite their limited capacity, authorities conduct basic health screening at the border, along with health services in the migrant reception and detention centres. Screening at *RHI-Haskovo* and *RHI-Sofia* showed prevalence of intestinal parasites (Giardiasis, Ascaridiasis, Blastocystosis) and our data indicates that the majority of refugees posed very limited infectious risk. Therefore, the findings confirm WHO

Table 4.2 Epidemiological data for 2016 – *RHI-Sofia*

Diagnosis	# of cases
Viral intestinal infection	2
Shigellosis	1
Enterocolitis	1
Varicella	1
Scarlet fever	1
Acute viral hepatitis E	2
Acute viral hepatitis B	1
Acute flaccid paralysis	1
Malaria	9
Giardiasis	71
Blastocystosis	42
Entamoeba coli	2
lodamoeba butschlii	2
Ascaridiasis	55
Trichocephalosis	3
Enterobiasis	2
Taeniasis	1
Total	**197**

Source: Created on the basis of the *RHI-Sofia* report

and DG SANTE expectations that migrants do not pose a risk to the local population.

At the same time, contrary to analysis by experts, public opinion reacted to the migrants situation as high-risk, demonstrated by the conflict in Harmanli migration RRS. Residents of the town of Harmanli protested on 20th November, demanding the camp's closure after local media reported on the suspected existence of communicable diseases on site: 'An artificially created tension led to this, following misleading reports that the centre is a hearth of infection,' Petya Parvanova, head of the Bulgarian Refugee Agency, was quoted by Reuters as saying.

The government declared that no medical reason for quarantine existed but the authorities took the decision to temporarily close the camp and restricted the free movement of migrants with the aim to calm down town citizens. In response, on 24th of November some 1 500-2 000 migrants (from the officially announced 3 070 registered in the camp) clashed with the police and the gendarmerie. As a result, most of asylum seekers have been moved to other centres. Meanwhile, on 18-19th November 2016, immediately after the first articles in the local media and first signs of discontent, a special medical mission from the Military Medical Academy was sent to verify the health status of the migrants concerned. The main findings of the military medical team study were as follows [16]:

- Very high share of patients with skin and infectious diseases: 128 patients with different symptoms;

Table 4.3 Refugee's health status, Harmanly RSS

Diagnosis	# patients	%	Notes
Pyoderma	36	28.1	
Scabies	26	20.3	
Varicella	18	14.1	Children exclusively from 1 to 8 years
Acute viral infection	15	11.7	
Dermatitis	16	12.5	
Others	16	12.5	Cystopyelitis, Candida vaginitis, neuro-vegetative dystonia, metrorrhagia, injuries, cuts and gunshot injuries
Cutaneous leishmaniasis	1	0.8	Suspected
Total	128	100	

Source: Created on the basis of the *MMA 2016* report [24]

- Very high level of male patients: 91 out of 128 total;
- Very high level of patients from Afghanistan: 106 out of 128 from Afghanistan;
- Usual pre-departure and journey-related health problems;
- Change in health profile after resettlement related to hygiene conditions;
- No emerging infectious and dangerous contagious diseases found.

As a result of the inspection, the military medical team has diagnosed the following diseases, ordered according to incidence (Table 4.3).

The conflict in Harmanly town illustrates some of the weaknesses reported in WHO's assessment of Bulgaria's capacity to manage a large migrant influx, notably risk communications and work with local media. Even though a special highly qualified medical team was sent to check the medical status of refugees and eventually to refute speculations of an epidemic, the conflict escalated instead of calming down. One explanation for this is the low levels of trust in the authorities and institutions in Bulgaria, which also includes lack of trust in healthcare professionals. In such circumstances, both clear risk communication and medical expertise are not sufficient to resolve a growing crisis.

4.4 Potential Risks to Public Health

In discussing the risks to public health, five topics are commonly considered: infectious diseases, vaccination, antimicrobial resistance, noncommunicable disease and bioterrorism (Fig. 4.12), as well as the linkages between them.

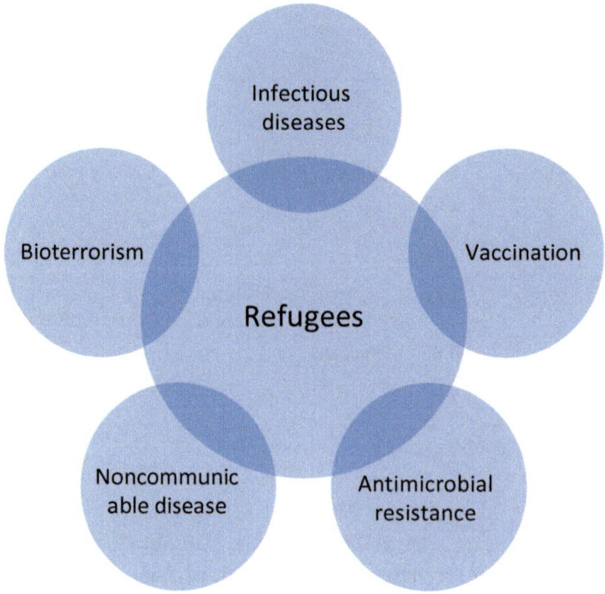

Fig. 4.12 Refugees as potential risks to public health

4.4.1 Infectious Diseases

With few exceptions, migrants are not at increased risk of transmitting communicable diseases [11]. However, infectious diseases can spread when new migrants live together in communal, close-quarter settings [18]. According to a recent survey in EU/EEA countries, screening for infectious diseases among migrants is currently directed towards predominantly human immunodeficiency virus (HIV), tuberculosis (TB), hepatitis B, hepatitis C, gonorrhoea, syphilis, measles and rubella, malaria and Chagas disease. These diseases were selected because the European Surveillance System (TESSy) collects data disaggregated by migrant status or because evidence suggests that they may disproportionately affect migrants in the EU/EEA [11]. Other diseases that could be considered include vaccine-preventable diseases, cholera, malaria, helminths and intestinal protozoa (Semenza et al. 2016) [21]. Epidemiological data from Bulgaria do not confirm the severity of TB, HIV and sexually transmitted diseases, but confirm the importance of the screening of parasitological diseases.

4.4.2 Vaccination

A potential risk to public health is the possibility of outbreaks of vaccine-preventable diseases (VPD) in a population coming from countries where immunization coverage is low [27]. Vaccinations to consider among migrants include: measles, poliomyelitis, meningococcal disease, and diphtheria/tetanus/pertussis. The risk of the spreading of vaccine-preventable diseases among the local population should not be underestimated too. The Varicella vaccination, which is not included in Bulgarian routine vaccination programs because the majority of individuals in temperate climates develop natural immunity from previous infection before adolescence, is a good example, especially when taking into account cases registered in the Harmanli refugee camp. Other possible threats could be recognized in the cholera epidemic in Iraq [2] and poliomyelitis in the Syrian/Lebanon border refugee camp [23]. Population displacement can also threaten global VPD eradication and elimination efforts [14].

4.4.3 Antimicrobial Resistance

Antimicrobial resistance (AMR) is not a disease in itself but a complication of the treatment of a disease. In situations such as the crowded settings with poor hygienic conditions found in refugee camps, infections can easily occur and spread; whether they are caused by resistant pathogens depends on their origin, which can be the environment, animals, food or humans [28]. AMR is becoming a global concern with AMR strains associated with new resistance mechanisms emerging and spreading worldwide [29].

The journey the refugees undertook, crowded conditions in refugee camps or settlements, and the lack of regular medical care, are prime drivers of the spread of AMR among this vulnerable group especially in multidrug resistance TB cases among refugees [5, 15].

4.4.4 Noncommunicable Diseases (NCD)

The range of potential risks that can be associated with refugees is inevitably wide-ranging. It includes communicable diseases, trauma, dermatological disease injuries associated with the journey, environment, changes in climate, especially during the winter, and last but not least, mental health and psychological problems [4]. WHO assessment in Bulgaria reported on physical and psychological trauma, the consequences of post-traumatic stress disorder; dehydration, nutrition disorders and hypothermia; absence or interruption of treatment for chronic diseases. Particularly high health risks are faced by vulnerable groups of migrants, including the elderly, people with disabilities, pregnant women and young children [27]. Last but not least, migrant's NCD treatment requires substantial resources that create additional economic pressure on the health systems of affected countries.

4.4.5 *Bioterrorism*

There are multiple causal relations between (forced/irregular) migration and terrorism – but these are generally complex [20]. While the link between terrorism and migration is widely discussed, where in the context of the Migration Inflow hypothesis immigrants are an important vehicle for the diffusion of terrorism from one country to another [3], the link between bioterrorism and migration is hardly found in official and scientific publications. However, bioterrorism as a risk and policy measures to address it is discussed in official documents of the Bulgarian government, particularly in the 2016 Annual Report of the Ministry of Health, which names as one of its aims to 'Protect the country from importation and distribution of infections with high epidemic risk by creating and maintaining mechanisms for timely and adequate response to health threats of a biological nature, including bioterrorism'.

The EU also includes bioterrorism in the list of new threats but not specifically in the framework of migration but in the general context of cross-border threats: 'Under EU law on cross-border health threats, existing mechanisms coordinate preparedness for serious cross-border threats to health, linking Member States, EU agencies and Scientific Committees through the Early Warning and Response System. The Health Security Committee, which coordinates Member States' responses to threats, may act as a focal point on vulnerabilities in public health, to enshrine hybrid threats, in particular bioterrorism [12]. Therefore, there is a wider recognition that migration and bioterrorism have to be dealt with in a broader institutional and policy context. Despite these general considerations of Bulgarian authorities and the EU Health Security Committee, none of our past experience, including the data from our research, convinces us that there is a clear link between bioterrorism and migration processes. Therefore, in the risk assessment, the likelihood of migrants being instrumental in a bioterrorist attack was assessed as **very low** (see Table 4.4).

Table 4.4 Evaluation of the potential risks to public health, Bulgaria 2017

Risks	Probability	Impact	Notes
No vaccinations	Very high	High	Measles, poliomyelitis
Antimicrobial resistance	High	High	
Spread of ID	Moderate	Moderate	
Non-communicable diseases	High	Moderate	Economic pressure
Bioterrorism	Very low	Very high	

4.5 Conclusion

Our findings confirm WHO and DG SANTE expectations that migrants do not pose a significant risk to the local population. At the same time some specific, health risks should be taken into consideration and categorised by varying degrees of probability and impact. The risk assessment should be made by a wide range of specialists based on detailed logitudinal data. Our assessment, based on the public data at our disposal, published and analysed in our article, is summarised in Table 4.4.

The assessment of these risks may seem overwhelming or unrealistic to the external observer, but it seems realistic when considering the geographical location of the country, the institutional weakness and the fragile public health capacity.

References

1. Alexander R (2017) Die Getriebenen: Merkel und die Flüchtlingspolitik: Report aus dem Innern der Macht. Siedler Verlag
2. Bagcchi S (2016) Cholera in Iraq strains the fragile state. Lancet Infect Dis 16:24–25
3. Bove V, Böhmelt T (2016). Does immigration induce terrorism? J Polit 78(2):572–588. doi:https://doi.org/10.1086/684679
4. Carballo M, Nerukar A (2001) Migration, refugees, and health risks. Emerg Infect Dis 7(3 Suppl):556–560
5. de Smalen AW, Ghorab H, Abd E, Ghany M, Hill-Cawthorne GA (2017) Refugees and antimicrobial resistance: a systematic review. Travel Med Infect Dis 15:23–28. https://doi.org/10.1016/j.tmaid.2016.12.001. Epub 2016 Dec 3. http://www.travelmedicinejournal.com/article/S1477-8939(16)30202-2/fulltext
6. EASO (2012) European Asylum Support Office. Annual report on the situation of Asylum in the European Union 2011 https://www.easo.europa.eu/sites/default/files/Easo_Annual_Report_2011.pdf https://doi.org/10.2847/15683
7. EASO (2013) European Asylum Support Office. Annual report on the situation of asylum in the European Union 2012. https://www.easo.europa.eu/sites/default/files/EASO_AnnualReport%202012.pdf. doi:https://doi.org/10.2847/38969
8. EASO (2014) European Asylum Support Office. Annual report on the situation of asylum in the European Union 2013. https://www.easo.europa.eu/sites/default/files/EASO%20Annual%20Report%202013.pdf. doi: https://doi.org/10.2847/28516
9. EASO (2015) European Asylum Support Office. Annual report on the situation of asylum in the European Union 2014. https://www.easo.europa.eu/sites/default/files/public/EASO-Annual-Report-2014.pdf doi:https://doi.org/10.2847/231335
10. EASO (November 2016) Latest asylum trends – November 2016. https://www.easo.europa.eu/sites/default/files/Latest%20Asylum%20Trends%20November%20draft%20wl.pdf
11. ECDC (2014) European Centre for Disease Prevention and Control. Assessing the burden of key infectious diseases among migrant populations in the EU/EEA. Stockholm: ECDC, 2014. doi: https://doi.org/10.2900/28792
12. EU (2016) European Commission. Joint communication to the EU parliament and the Council. Joint Framework on countering hybrid threats a European Union response. Brussels, 6.4.2016, JOIN (2016) 18 final. http://eur-lex.europa.eu/legal-content/EN/TXT/?uri=CELEX%3A52016JC0018

13. Jakab Z (2015) Population movement is a challenge for refugees and migrants as well as for the receiving population. http://www.euro.who.int/en/health-topics/health-determinants/migration-and-health/news/news/2015/09/population-movement-is-a-challenge-for-refugees-and-migrants-as-well-as-for-the-receiving-population
14. Lam E, McCarthy D, Brennan M (2015) Vaccine-preventable diseases in humanitarian emergencies among refugee and internally-displaced populations. Hum Vaccin Immunother 11(11):2015
15. Mihaylova-Garnizova R, Plochev K (2013) Health case study – Bulgaria. In: Hunger R, Belojevic R (eds) Biopreparedness and public health, 1st edn. Springer, pp 75–89
16. MMA (2016) Military medical academy. http://www.vma.bg/bg/novini/spetsialisti-ot-vma-ne-otkriha-masova-zaraza-sred-bezhantsite-v-tsentara-vharmanli-193.html
17. Morabia A, Benjamin GC (2015) The refugee crisis in the middle east and public health. Am J Public Health 105(12):2405–2406. https://doi.org/10.2105/AJPH.2015.302929
18. Permanand G, Krasnik A, Kluge H, McKee M (2016) Europe's migration challenges: mounting an effective health system response. Eur J Pub Health 26(1):3–4. https://doi.org/10.1093/eurpub/ckv249
19. SAR (2017) State Agency for Refugees with the Council of Ministers, Bulgaria. http://aref.government.bg/?cat=21. Accessed 1 Apr 2017
20. Schmid AP (2016) Links between terrorism and migration: an Exploration, ICCT Research Paper. DOI: https://doi.org/10.19165/2016.1.04
21. Semenza JC, Carrillo-Santisteve P, Zeller H, Sandgren A et al (2016) Public health needs of migrants, refugees and asylum seekers in Europe, 2015: infectious disease aspects. Eur J Pub Health 26(3):372–373. https://doi.org/10.1093/eurpub/ckw023
22. Seychell M (2016) European parliament, Briefing. Martin Seychell, Deputy Director General of the European Commission Directorate for Health and Food Safety, The public health dimension of the European migrant crisis, January 2016. http://www.europarl.europa.eu/RegData/etudes/BRIE/2016/573908/EPRS_BRI(2016)573908_EN.pdf
23. Sharara SL, Kanj SS (2014) War and infectious diseases: challenges of the Syrian civil war. PLoS Pathog 10(11):e1004438. https://doi.org/10.1371/journal.ppat.1004438
24. UNHCR (2016a) UN Agency for refugees. Winter Operations Cell – Daily Report, 9 March 2016: http://data.unhcr.org/mediterranean/download.php?id=840; https://data2.unhcr.org/fr/documents/download/47170
25. UNHCR – UN Agency for refugees (2014) Asylum trends 2013: levels and trends in industrialized countries. Office of the United Nations High Commissioner for Refugees, Geneva
26. UNHCR – UN Agency for refugees (2015). World at war: UNHCR global trends: forced displacement in 2014. http://www.unhcr.org/556725e69.html. Assessed 15 Jan 2017
27. WHO (2015a). Bulgaria: assessing health-system capacity to manage sudden large influxes of migrants. Joint report on a mission of the Ministry of Health of Bulgaria and the WHO Regional Office for Europe. http://www.euro.who.int/__data/assets/pdf_file/0009/300402/Bulgaria-Assessment-Report-en.pdf
28. WHO (2016a). Migration and health: key issues. http://www.euro.who.int/en/health-topics/health-determinants/migration-andhealth/migrant-health-in-the-european-region/migration-andhealth-key-issues#292117. Accessed 9 May 2016
29. WHO (2016b). Antimicrobial resistance. http://www.who.int/mediacentre/factsheets/fs194/en/. Accessed 28 Sept 2016

Chapter 5
A Perspective on the Strategy of Intelligence

Randall Murch

> *If you know the enemy and know yourself, you need not fear the result of a hundred battles. If you know yourself but not the enemy, for every victory gained you will also suffer a defeat. If you know neither the enemy nor yourself, you will succumb in every battle.* Sun Tzu (Chinese General, 6th Century BCE), from *The Art of War [1]*

Abstract In writings on and discussions of bioterrorism prevention and control over the past two-plus decades, the subject of intelligence has received limited attention. This chapter is intended to provide a foundation for either scholars or others newly interested, and government and military program managers and decision makers interested in establishing a new program or developing a fledgling one. It could also be used to catalyze tailored studies, as well. A broader approach to this topic has been used to provide as much information as possible to not only inform and educate, but enable further analysis, discussion and action which the reader can tailor to their own respective interests, further research, goals, objectives and outcomes. It is not written to direct the reader on a specific path or to a specific end point. The objectives of this chapter are: provide a broad foundation of intelligence and its value to countering bioterrorism and bioproliferation; identify some critical challenges; describe how national security and law enforcement leverage each other and other organizations which have necessary expertise, resources, authorities and information; frame and illustrate intelligence gathering, analysis and use to counter bioterrorism or bioproliferation through examples and plausible scenarios; to stimulate thought, questions, discussion and engagement as one looks at one's own national or collaborative multilateral programs; and, introduce possible "next steps",

R. Murch (✉)
Virginia Polytechnic Institute and State University (Virginia Tech), Arlington, VA, USA
e-mail: rmurch@vt.edu

V. Radosavljevic et al. (eds.), *Defence Against Bioterrorism*, NATO Science for Peace and Security Series A: Chemistry and Biology,
https://doi.org/10.1007/978-94-024-1263-5_5

which could include NATO assuming leadership to craft and guide follow-up engagements specifically on this topic.

5.1 Introduction, Objectives and Background

The term "intelligence" is used in many venues and contexts, for not only homeland or national security, but also even in business. Here, it will be discussed with national and global security in mind, with a focus on bioterrorism prevention and control in particular.

Intelligence collection, analysis, use and improvement is a complex system of systems, there is no way around this. Regardless of the investments made and resources stood up, intelligence faces many challenges and uncertainties; it is the "nature of the beast". For a number of reasons, effective intelligence against Weapons of Mass Destruction (WMD) programs, small to large and crude to sophisticated, is particularly difficult. Those who develop, acquire, test and prepare to use WMD usually keep critical information about their capabilities and intent to need to know by only a select few and configure them for protection, in order to sustain programs, enable surprise and achieve desired outcomes. From my experiences and encounters with many experts, it is clear that biological weapons activities present unique uncertainties and complexities not possessed by other WMD activities, making it even more challenging. This chapter is intended to provide a treatment of this realm, particularly for those readers who have not considered it as yet.

Defense against bioterrorism, with the goals or prevention and control, requires that competent, standing, adaptive, rapidly configurable and deployable capabilities exist *a priori* for best possible performance and value. These are in addition to and integrated with effective intelligence collection, analysis and use. Such capabilities include: public health and agricultural biosurveillance, epidemiology and outbreak response; environmental and hazardous materials management and response; proper management and regulation of "dual use" science to limit nefarious and illicit purposes and activities; law enforcement that has the appropriate authorities, trained personnel and resources which can be adaptively and dynamically applied from "trigger" (initial awareness that an activity of interest is occurring) through attribution for either intervention to reduce risk or investigations to support prosecutions; laws and a legal system that can effectively address prosecutions and exonerations; information sharing among components charged with supporting defense, anticipation, characterization, response and resolution across the enterprise; integrated multidisciplinary analysis and situational awareness; timely and effective decision making at the tactical, operational and strategic levels; and, scientific and technical reachback that addresses the requirements and gaps of the aforementioned domains.

Many of the aforementioned have received attention in publications and public presentations over the past 20 years. **Intelligence** is often overlooked and will be developed in this chapter from a strategic perspective (Fig. 5.1).

- ❖ Intelligence to counter the biological threat may not be a priority presently, but in alliances and countries where it is not, perhaps the case should be made that it needs to be, even for low probability, high consequence events
- ❖ Intelligence is a critical aspect of counter-bioterrorism and counter-bioproliferation programs and enterprises
- ❖ Effective and timely intelligence is necessary to inform decisions at all phases and levels
- ❖ Intelligence is a dynamic, not static process and national capabilities should be implemented with this in mind
- ❖ Effective intelligence requires planning and sustained commitment and investment
- ❖ Perfect intelligence is difficult, if not impossible to achieve, and the cost of a large enterprise may be too burdensome; however, some well-placed intelligence sources and methods could provide informed and timely situational awareness, and inform assessments of threats and risk
- ❖ A structured and coordinated intelligence enterprise is far superior than a stochastic and disorganized one
- ❖ One is never "done" seeking sufficient intelligence or intelligence advantage on current, emerging and future threats

Fig. 5.1 Strategic considerations for creating or enhancing intelligence capabilities focused on bioterrorism, biowarfare and bioproliferation

The objectives for this chapter are:

- Provide a High-level and Broad Perspective of Intelligence and Its Value to Countering Bioterrorism and Bioproliferation
  - Intelligence is necessary to avoid surprise through indications and warnings, aiding in mitigating threats through early interventions, supporting effective response, assisting with attribution which, in turn, supports decisions leading to actions through various means
- Provide Understanding Regarding What Intelligence Assets Exist and How They Usually Function and are Organized
  - Intelligence is conducted by government organizations through a variety of sources and methods. Although these may be known in general, they are necessarily protected with regard to what exactly these entail and how, when, where and for what purposes they are applied to meet operational, analytic and decision making requirements.
- Identify Critical Challenges
  - Effective and timely intelligence is a difficult undertaking, and that supporting protection against and control of bioterrorism is one of the more difficult areas for which it exists, or should
- Describe How National Security and Law Enforcement Leverage Other Agencies Which Have Necessary Expertise, Resources, Authorities and Information

  - In countries that have prioritized preventing and countering bioterrorism and related nefarious and illicit endeavors, structures, processes and mechanisms exist by which national security and law enforcement avail themselves of required capabilities and personnel, particularly in required technical specialties, which do not exist or are limited in their respective organizations.

- Frame and Illustrate Intelligence Gathering, Analysis and Use in the Contexts of Countering Bioterrorism or Bioproliferation Through Examples and Plausible Scenarios

  - Plausible and well-crafted scenarios are widely used by military, intelligence, law enforcement and homeland security agencies as tools to enhance education, training and exercises, planning and preparedness, analysis, and vision and strategy development. Their use helps to ensure and enhance agency and cross-agency coordination, alignments of capabilities, and understanding requirements, needs and gaps and how to plan for and address these.

- Inform Those Who Do Not Have Expertise or Experience in These Matters, But Are Either Interested or in Positions from Which They Could Offer Assistance or Provide Support to Security or Law Enforcement Agencies in Their Respective Countries

  - Even the best organized and resourced emergency response capabilities will not prevent, or acceptably mitigate, a bioterrorism incident or the proliferation of biological weapons or source materials. Further, science or public health together or alone are not sufficient.
  - Effective and timely intelligence is necessary, as is law enforcement. A "systems approach" is warranted and prudent. To gain a fuller picture, and potentially understand how one might engage, this chapter could provide those from other domains basic knowledge with which to reach out to their security services or law enforcement, or be willing to engage should the latter do so.

- To Stimulate Thought, Questions and Discussion to Address in the Moment, and Perhaps to Take Away for Subsequent Consideration and Action

  - I am assuming that some readers could creatively influence how their governments prioritize, staff, resource and implement counter-bioterrorism intelligence plans, programs and capabilities. If intelligence is not taken seriously or is not treated as a sufficiently high priority, risk significantly increases. Even one leader-innovator or a small number can make a difference.

Questions to Ponder

- Does your country and its national security – law enforcement enterprise prioritize bioterrorism (bioproliferation) as a threat or concern?
- What intelligence capabilities does your country have which are directed at or are inclusive of bioterrorism and bioproliferation? If they do not exist, should they be created? If they do, can or should they be improved?

- How does NATO treat bioterrorism and bioproliferation within the alliance, with regard to intelligence production and intelligence sharing? Should collaboration exist or be improved?
- Given that many intelligence resources, methods and applications will necessarily be hidden from view, can any "open source" (public) initiatives be developed and implemented in the near term which could produce beneficial outcomes?
- What is the relationship between the intelligence, security and law enforcement, emergency services, and the scientific and public and agricultural health communities in your country or within the NATO community? Can these be configured to better leverage each other for improved, integrated bioterrorism and bioproliferation prevention, anticipation and risk reduction?
- If intelligence focused on bioterrorism and bioproliferation is non-existent or a low priority in your country or within the NATO partnership, should this be changed? What can you do to help change this?
- If intelligence is or will not be treated as a priority, is "better lucky than good" an acceptable or prudent approach to effectively address high-impact threats and outcomes such as those that bioterrorism would manifest? What can be done to change this mindset?

5.2 Providing a Baseline for the Strategy of Intelligence

Let us begin with terms of reference (definitions). For our purposes, **Intelligence** can be defined as the collection, analysis, integration and use of information of military, national security, policy or political value [2]. Some criminal investigative agencies also generate what they refer to as intelligence for their investigative purposes or to share with other agencies and use some of the methods and means discussed below.

Intelligence is undertaken:

- to provide indications and warnings on actual or possible situations, events and phenomena of interest or concern
- to characterize and assess key individuals, groups or enterprises of interest or concern
- to understand strengths, attributes, capabilities, resources, limitations, weaknesses and vulnerabilities of adversaries or potential adversaries
- to understand, anticipate and address current, emerging and unknown threats
- to inform future, near-term and real-time critical decisions for interventions and response at the tactical, operational and strategic levels
- to develop strategies, plans, preparations, courses of action, and operations or other activities which increase likelihood of success while minimizing risk

There are two general categories of intelligence, Positive Intelligence and Counterintelligence, which the author will seek to define and illustrate to make these concepts more understandable than when articulated by scholars of intelligence [2, 3].

Positive Intelligence Collection of, analysis and use of intelligence against individuals, groups, organizations, nations, policies, strategies, plans, situations, facilities, capabilities, phenomena, threats, decision making and technologies (which adversaries or those of interest use, or show interest in or are pursuing).

Plausible, Hypothetical Example 1: *Collection and analysis against a foreign leader of strategic political interest*

There has been an unexpected result in the presidential election of a country which shares a border and has strong ethnic, cultural, trade and security ties with its neighbors. The winner has "come out of nowhere" and stunned the incumbent, the media which has been monitoring and reporting on the election, and in-country academic experts and those from institutions outside the country. The articulated priorities of the new president seem to be a radical departure from what has existed in that country and driven relations between its neighbors for the past 18 years. But, since the new president has not yet been inaugurated nor actually begun to populate her political appointment and begin to govern, it is too early to tell what paths she will take. To ensure that each of the neighboring countries best position themselves to understand the new leader, prepare to engage her and her administration and advance their respective interests, the president or prime minister of each one directs the respective intelligence services to collect the appropriate intelligence and provide their assessments on a weekly basis until directed otherwise. There are trusted relationships between some of the neighboring country intelligence services, so intelligence sharing is expected. The services focus on the new president's origins, education, career, public communications and personal and professional networks, first through OSINT (open source intelligence). Once an initial knowledge base has been created and OSINT is continuing, the services move to refine and provide continual updates through the use of HUMINT (human source and human-enabled intelligence). Individuals, for which their affiliations with the respective intelligence services are unknown except to those services, are tasked with specific collection missions specifically focusing on policy, economics, trade and security intentions of the new president with the neighbors. Two of the neighbors are concerned that other countries that may attempt new or stronger ties with that country which would result in a large and standing presence which would be of substantial concern. The senior leadership of two neighboring countries have ordered communications intercepts (SIGINT, signals intelligence) on the personal mobile phones of the new president, her closest advisors and the candidates for minister of foreign affairs, economy and trade, and internal and border security and to expand this coverage to office communication systems following her inauguration.

Plausible, Hypothetical Example 2: *Collection and analysis against a suspected bioproliferator*

A friendly foreign intelligence service in Europe has contacted U.S. Intelligence and provided information that a previously unknown individual, representing herself as a principal in a new start-up venture capital firm, who recently approached a well-known, established pharmaceutical company headquartered in its country. It was just announced that the company is an awardee of a U.S. Government biodefense contract for a new therapeutic intended for U.S. military forces which have been exposed to certain endemic infectious disease agents in Southeast Asia. Because of the manner in which this individual presented herself and the questions she was asking, the company became suspicious and had their security department perform an initial check on the identifying information that was provided on the firm and the contact information provided, using various resources available including their national law enforcement agencies. The friendly foreign agency is asking U.S. Intelligence for assistance in particular with its vast array of signals intelligence (SIGINT) and complex data analytics capabilities to determine true identity of this individual, actual country of origin, affiliations and funding sources. U.S. Intelligence agrees to this request. U.S. Intelligence is successful and identifies this individual as being in the employ of a foreign intelligence service of a country that considers the U.S. and which is suspected of having a nascent biological weapons program. Given that both the requesting service and U.S. Intelligence have equities in this matter. They agree to conduct and coordinate intelligence operations against this individual and the organization she is employed by. Joint intelligence mission planning has been instituted against her and the employing intelligence service.

Counterintelligence Intelligence gathering and analysis which is focused on protecting against adversaries' or others' access to or exploitation of the above, or unwarranted and unauthorized access or release of critical and sensitive information, which could include strategies, plans, operations, technologies, resources, capabilities and vulnerabilities (see Refs. [3, 4]).

Plausible, Hypothetical Example 1: *"False Flag" operation against would-be purveyor of sensitive information*

The U.S. Federal Bureau of Investigation (FBI) obtains human source intelligence (HUMINT) reporting through one of its field offices which identifies suspicious activities of a recently retired U.S. Army officer who has been hired as a technical program manager at a local company which is funded by U.S. Government contracts. This individual is seeking to sell sensitive technical performance and specifications information on a new generation of battlefield intelligence, surveillance and reconnaissance systems being developed by his new employer. Exposure of these systems would put U.S. troops and operations at high risk. The FBI opens a counterintelligence investigation on this individual and obtains his military service record, performs record checks and locates his residence and institutes physical surveillance to understand his daily activity patterns and associations (places and persons). Access to his financial records suggests that he cannot meet his financial obligations, which began before he left the military, for various reasons. The FBI chooses to ensnare this individual and intervene in such a way to eliminate the risks posed by conducting a "false flag" operation. That is, an undercover FBI Special Agent will pose as a foreign intermediary who brokers deals such as this for a foreign country which is most interested in acquiring such information. The FBI completely choreographs the meetings, chooses meeting locations with lavish accoutrements, and ensures that the undercover Agent has all of the backstopping needed to ensure that the "target" completely believes that the Agent is who he says he is. Once the "trust" is developed, the target becomes more aggressive with offers of sensitive information that he can access and provide, if payments will match the value. The undercover Agent makes a detailed request and insists on a deadline but offers a substantial bonus if the target meets both. Instead of promises and insinuations as has occurred up to this point, at a meeting in a very secluded location, the target provides the most sensitive information available on these systems (which has been artfully and believably counterfeited) in exchange for the agreed upon sum and a 40% bonus. Immediately after the exchange occurs, the target and "intermediary" are arrested by FBI Agents hidden in close proximity. The target is subsequently indicted, prosecuted, convicted and imprisoned. The "intermediary" returns to his field office and routine duties and awaits his next assignment.

Plausible, Hypothetical Example 2: *Joint Interagency Task Force counterterrorism operations against group planning penetration of a laboratory for access to stock cultures to prepare for a bioterrorism attack*

Due to a national university biosecurity engagement and awareness program that is conducted by the National Security Service (NSS) and which recently visited his university, the director of a prominent laboratory which conducts research on human and zoonotic pathogens has requested a meeting with local NSS field office personnel for more detailed discussions. (This lab has fully-outfitted BSL-3 spaces.) Soon after the request, the professor and his senior staff meet with the local NSS. During this meeting, one of the senior associates mentions that she has observed that one of the foreign postdoctoral fellows in the lab has been showing more than usual interest in experiments with specific pathogens that are occurring in one of the BSL-3 labs. That postdoc has been asking detailed questions of other lab personnel about the pre-publication results of the experiments, how the pathogen could be manipulated and whether these experiments would violate national law, who exactly has control of the stock cultures and the collaborators in other universities. That postdoc is not on the team assigned to this work. Further, the postdoc has approached two of the technicians in that program asking about details of the stock cultures. When the postdoc is engaging others, he has mentioned that his home country is interested in the research for its own public health purposes and priorities and "he is just trying to help". The lab has received no official or unofficial inquiries from the public health ministry or prominent universities in that country. The NSS decides to open a preliminary inquiry on this postdoc to ascertain the nature of intentions, plans, activities and associations.

Fig. 5.2 Major categories of intelligence sources and methods

SIGINT: Signals Intelligence

HUMINT: Human Intelligence

OSINT: Open Source Intelligence

GEOINT: Geospatial Intelligence

IMINT: Imagery Intelligence

Intelligence Sources and Methods This refers to the array of means, approaches, techniques, capabilities, procedures, tactics and methods that are used or developed to accomplish intelligence gathering, which feeds analysis and use. For the most part, these exist, are developed and applied in clandestine ways (see Refs. [5–7]). These are categorized in various ways, but the text box and following brief discussion should provide a useful summary (Fig. 5.2).

Signals Intelligence (SIGINT): SIGINT is further broken out as either Communications Intelligence (COMINT; e.g., voice or data collection, intercept or tracking) or ELINT (Electronics Intelligence; which includes targeting signals that do not include speech or text, e.g., telemetry data from enemy missiles or emissions from shipborne anti-threat systems). SIGINT collection can occur in tactical, operation and strategic constructs and situations and occurs via many types of land, sea, air and space platforms. A microphone concealed in proximity to those having a conversation which is expected to produce valuable intelligence is an example of tactical SIGINT. Tracking of mobile phone GPS data for one or more targets through one or more service providers to pinpoint locations and characterize activities over time, which leads to the collection of specific mobile text and call metadata, providing for associative, organizational network and timeline analysis is an example of operational-level SIGINT. Strategic-level SIGINT could be considered as massive, highly sophisticated collection and analysis of bulk voice and data communication of a target through complex, multilevel systems. Another example of strategic SIGINT is collection and analysis which targets electronic communication sent from, received by and emissions emanating from a third party aircraft carrier battle group which is repositioning in response to specific pronouncements and actions of protagonists and antagonists in a geographical region of strategic importance.

Measurement and Signature Intelligence (MASINT): This intelligence sources and methods domain is constituted by a highly technical array of signature detection and measurement systems and procedures which include electro-optical (including spectroscopic, hyperspectral and laser), nuclear radiation, geophysical (weather, acoustic and seismic), radar (e.g., synthetic aperture radar), materials (nuclear test, radiological, biological and chemical) and radiofrequency (e.g., non-cooperative target recognition) means supported by an array of technologies [6]. Space limits a full discussion of these but for the topic at hand, "Bio-MASINT" would be the most commonly applied. This could take the form of clandestine collection or detection and subsequent analysis of emissions or waste coming from a suspect facility or the forensic analysis and possible attribution and association of samples taken from both a suspected bioweapons laboratory and test sites and associated dead test animals.

Human-Source Intelligence (HUMINT): Human source intelligence is gathered or collected by human operatives who are in direct or indirect contact with those they seek to acquire specific information about, or of value from or through. A variety of approaches are used, to include what would be commonly known as espionage, the operative directly collecting it, or directing someone to acquire specific and sensitive information or accepting it if offered and determined to be from a valuable or trusted source. Human intelligence can also be acquired through interviews or interrogations of persons with knowledge or experience of value, such as those who have traveled to a country or region of interest or those who have been able to directly observe or have had contact with a person or place of intelligence interest. HUMINT can have tactical, operational or strategic value and is categorized depending upon the value and impact it has and who the intended user is.

Open-Source Intelligence (OSINT): This is gathered from a variety of open (unprotected, unclassified) sources, such as news media, press conferences, scholarly journals, scientific and technical publications, speeches and presentations, a wide variety of Internet sites and sources, and public meetings, trade shows and conferences and publications produced from such and company publications. While it is well known that intelligence services access OSINT and some publish open reports from it [8, 9], it is also true that its specific uses can be sensitive or classified.

Geospatial Intelligence (GEOINT): Geospatial intelligence is gathered using satellite and aerial platforms and includes both images and mapping/terrain data. Imagery Intelligence (IMINT) is either a category or subcategory that usually refers to imagery that is gathered from satellite and aerial photography.

Other categories of intelligence gathering methods that are sometimes mentioned [7]:

Technical intelligence, or TECHINT, which gathered from analysis of weapons and equipment used by the armed forces of foreign nations, or environmental conditions.

Medical intelligence (MEDINT) is gathered from analysis of medical records, examinations to determine health status, particular ailments, and specific conditions of interest or health reporting in particular populations in specific locations.

Cyber Intelligence (CYBINT) and Digital Network Intelligence (DNINT) are gathered from cyberspace.

FININT (Financial intelligence) is gathered from analysis of monetary transactions.

The reader should not get too caught up in the categorization or labeling of intelligence gathering methods. Rather, just realize that there are major types, many specific technologies, technical systems and human skill sets employed, and a considerable array of methods, tactics, techniques and procedures used (also known as "tradecraft"). Intelligence gathering is tailored to the desired or intended outcome and is a dynamic process for both individual and standing requirements.

The uses of INTs, singly or in combination, is chosen and tailored depending on the intelligence requirement (users stated need or tasking). How collection is tailored depends upon a number of factors pertaining to the individuals or organization to be targeted, their patterns, associations and business cycle, frequented locations, communications methods and means, operational and communications security, operational sites and environments, procurement of key materials and equipment, supply chain and logistics, and transportation modalities. For suspected or identified bioterrorists, bioweaponeers, bioproliferators, enablers, suppliers, logisticians and commanders and controllers, intelligence services could apply variants of SIGINT (CYBINT, DNINT and FININT), HUMINT, MASINT, OSINT and even TECHINT (MEDINT) with associated analytic capabilities to address specific, general or even emerging intelligence requirements.

Protection of Sources and Methods Intelligence sources and the methods used to collect and analyze it, as well as results, conclusions, decisions, actions taken and successes and failures are most often protected from adversaries, those who support them, potential adversaries and the public. (See at Ref. [10]) This is for several reasons: (1) intelligence methods, supporting technologies and infrastructure are often costly and risky, and take considerable time, effort and resources to develop, deploy and maintain effectiveness; (2) intelligence methods are likely to be successful if the target is unaware or poorly informed or mistaken with regard to their existence, uses, variants, strengths and weaknesses; (3) once the specifics of methods are exposed or known, even within a "closed system" of intelligence and counterintelligence, adversaries can learn from discoveries, and develop countermeasures can be which prevent or reduce successful collection, or improvements can be made and used for their own collections; (4) sometimes, the effectiveness of intelligence collection and analysis capabilities can be correlated to what a target believes to be true, rather than what is true in fact; and (5) exposure of sources and methods to the public via OSINT (including news reporting) also enables adversaries to learn and respond with countermeasures rendering them potentially less effective or ineffective, or requiring additional resources to "re-engineer" them or create *de novo*.

The Intelligence Cycle This ecosystem, which encompasses the tasking, collection, analysis, and delivery of intelligence, is organized in a cyclical process, as depicted in the figure below (Fig. 5.3).

This simple diagram indicates the major steps and relationships of those steps to one another in what is commonly termed, The Intelligence Cycle. However, a single, simple diagram cannot capture what each step is comprised of, the human and resource capital involved and the nuances and subtleties contained in each step and the overall cycle. Further, when there are more than one agencies or department involved, and multiple sets of requirements are being addressed, even against the one target, a complex choreography occurs and must be properly and deftly managed

Fig. 5.3 Two views of the intelligence cycle: US Federal Bureau of Investigation and New Zealand Security Intelligence Service [11, 12]

to ensure the requirements are prioritized and addressed, users receive the best possible, most timely intelligence and information to inform decisions and actions, conflicts are adjudicated, problems are resolved or satisfactorily navigated, risks are minimized and managed, sources and methods are protected and operational security is maintained. The bigger the intelligence enterprise, the more complex the cycle.

The general process and functioning of the cycle will be illustrated in the context of bioterrorism with a plausible hypothetical scenario.

The internal security service of a moderate Middle Eastern country becomes aware of conversations between radical, anti-regime elements that they have under routine surveillance and an extremist sub-state group which has elements in two neighboring nations but not their own. These communications are interpreted as the former seeking support from the latter to acquire and use biological weapons against targets within the country. This is the first time the internal service has had any information that bioterrorism could be a threat to national security and public safety. The internal service communicates this intelligence with the national police agency and the country's external intelligence service. Neither have any information on this group's or external group's interest in bioterrorism. The three agencies agree to cooperate on vital intelligence gathering on these two groups specifically with regard to bioterrorism and associated activities. They begin by determining the intelligence priorities, who will be responsible for addressing particular requirements and what the analysis will be expected to produce. They organize using a tailored intelligence cycle which leverages the missions, authorities and available resources in the three agencies. All three realize that they will require technical expertise and resources that do not reside within the respective agencies. They decide to task initially task HUMINT and SIGINT against the leaders of the internal and external extremist groups and someone identified as a confidant. Within a few weeks, based on SIGINT and HUMINT, a confidential source is developed who is in position to provide valuable information on strong intent and limited capabilities of the internal group. The external service successfully targets mobile phone and computer communication and produces high quality intelligence which strongly suggestive that the internal extremist group is seeking biological weapons and equipment to execute biological attacks but does not have the expertise to plan and execute such. SIGINT collections on the "Chief Scientist" of the external group indicates that they have access to what is needed and have agreed to provide it for a price. The two are monitored negotiating the specifics. Based on what has been obtained so far, the three agencies agree to raise the priority of intelligence targeting against the two groups, and to expand SIGINT to include financial transactions. During the next cycle of collection, analysts realize that the internal group has chosen a particular infectious disease organism to use as their primary weapon in small-scale airborne releases and another through a foodborne release. No targets are mentioned as yet. An inquiry within the police and security services determines that no expertise exists on these pathogens or what the effects and outcomes of using such would produce. They quietly approach their Ministry of Health, who in turn, identifies leading experts at the top university in the country who they work with closely and could be approached to provide technical support. The internal and external services have limited MASINT capabilities and communicate requests for assistance through standing relationships that they have with friendly foreign services which do possess such. Intelligence operations are expanded to position and use MASINT resources which are focused on detection and characterization at a site that the HUMINT source has identified as a likely makeshift laboratory. MASINT microbial forensics and attribution expertise and technologies are made available by the foreign services. The laboratory is now prioritized as a SIGINT target. IMINT is attempted on that site but no useful information is obtained and is discontinued. Soon thereafter, one of the friendly foreign services provides intelligence on two companies which they have been aware of that have staff who are communicated with the transnational extremist group using personal email accounts. The content of the communications is undergoing further analysis to determine its nature and value. The Middle Eastern country's internal service determines that the transnational group is sending a PhD microbiologist to meet with the internal extremist group. Partial identifying information is obtained on this individual, but not sufficient to pinpoint who this individual is and when or how he or she might be traveling or where the meeting will take place. The confidential HUMINT source is securely tasked to determine the identity of this individual and travel meeting particulars. After 4 days, the report back from that source is still pending, though understandable given the operational security that source and the service must use to avoid detection. Meanwhile, the Ministry of Health assists the internal security service and national police with recruiting the assistance of two leading university scientific experts and the promise by them to recruit additional expert colleagues if necessary. These experts are familiar with microbial forensics and legal requirements of the science for prosecution and claim their analytic methods would meet these even though microbial forensic has not been introduced at trial to this point. Undercover physical surveillance units of the national police detect increased activity at the makeshift lab site. The internal security service detects an upsurge in mobile phone traffic by key target individuals which is indicates that "an important visitor is arriving and operational planning must begin". Physical surveillance also reports that several deliveries have been made by commercial vehicles at the makeshift site which are in addition to normal traffic. Based on the analysis of daily patterns at the site, SIGINT collections, confirmation of the identity and presence of the PhD microbiologist in question (HUMINT source), a covert joint national police – internal service technical surveillance team conceals a camera and microphones within the facility to record activities and conversations. The leadership of the three services have consulted the Ministry of Justice and the Office of the Prime Minister which have agreed that if "overt acts" are determined from these methods, taken with all of the other evidence and intelligence that could be revealed in open court, a "takedown" will be authorized. It is anticipated that some of the intelligence, sources and methods will not be revealed to avoid compromise and will be used by the external intelligence service to directly or indirectly pursue and neutralize the external extremist group. The site compromise methods are successful. As a result, one week later the Prime Minister's Office, in consultation with the Minister of Justice and National Security Advisor, orders raids on the makeshift lab, and residences and other sites used by the internal extremist group. Incontrovertible evidence as well as highly valuable intelligence is expected from collections from all sites, including what appears to be the two weaponized pathogens and fabricated dissemination devices. The planned bioterrorism missions have been foiled. Those arrested are taken into custody by the national police. Interrogations will follow, as will exploitation of evidence and intelligence gathered from the raids of the various extremist sites. All of this could inform follow-on police and internal and external security service actions. Soon thereafter, the leadership of the three organizations involved order a review for "lessons learned" and planning for improved intelligence, surveillance and technical capabilities against bioterrorism and bioproliferation.

5.3 Deeper Looks with Bioterrorism Prevention and Control in Mind

5.3.1 Agencies That Are Best Positioned to Produce Intelligence Related to Bioterrorism Prevention and Control and Others Which Can Support

Since I am an American, I will use the U.S. Intelligence Community as a framework for this part of the discussion. Each reader is invited to examine how their own nation's intelligence enterprise is organized, how respective agency missions and responsibilities are delineated, and how each interacts with others to plan, execute against priorities, react in the face of exigencies or take advantage of unexpected

Fig. 5.4 High-level structure of the US Intelligence Community. Agency Designations: FBI is the Federal Bureau of Investigation, DEA is the Drug Enforcement Administration, Energy is the Department of Energy, DHS is the Department of Homeland Security, State is the Department of State, and Treasury is Department of the Treasury [13].

opportunities. Then, factor in bioterrorism prevention and control, and assess the likelihood of success or failure when considering the associated complexities and uncertainties with bioterrorism and the intelligence resources and capabilities possessed or needed to anticipate, mitigate or resolve them. The reader can take the next step and consider how their respective agencies and capabilities can be adapted to improve "biointelligence" within the various constraints that exist.

The U.S. Intelligence Community is a vast and complex enterprise which has innumerable organizational units, resources, capabilities, requirements, methods, processes and mechanisms which constitute it. A diagram depicting the functional and relational aspects of the U.S. Intelligence Community would be overwhelmingly complex. Something of this size and magnitude may not exist in every country, but the author warrants that, no matter, the principles and brief discussion for each set forth below can be used as "lenses" through which to analyze whatever nation's intelligence apparatus one chooses (Fig. 5.4).

The U.S. Intelligence Community is comprised of 17 agencies from both civilian and military sectors. From the civilian side, these include the Central Intelligence Agency (independent agency), the Federal Bureau of Investigation (Department of Justice), Department of Homeland Security, Department of Energy and Department of the Treasury. Not all of the organizational components, resources and capabilities of these agencies are devoted to intelligence but some are. From the military side, specific intelligence branches and components exist in the services (Army, Navy, Air Force and Marine Corps) with agencies being specifically devoted to intelligence or intelligence support (those that support the entire Department of Defense and others, specifically the National Security Agency, Defense Intelligence Agency, National Reconnaissance Office and National Geospatial Intelligence Agency). The Coast Guard is part of the Department of Homeland Security. The Office of the Director of National Intelligence (ODNI) is the coordinating and overarching policy setting entity for the U.S. Intelligence enterprise. Specific offices within the ODNI and some agencies are directly interested in bioterrorism, biowarfare, bioproliferation and the misuse of the life sciences.

This chapter is not intended to be a detailed treatment of the U.S. Intelligence of any nation's intelligence community. However, in addition to understanding how any intelligence enterprise and its components might function, I provide illustrations or discussion regarding the purpose of intelligence as in the context of bioterrorism, bioproliferation and biowarfare prevention and control:

5.3.2 *Agencies Have Differing Missions, Authorities, Responsibilities, Resources, Priorities and Customers*

Due to its requirements, a military service seeks intelligence on foreign programs that might be developing and intending to use biological weapons (biowarfare) to ensure that they have effective medical countermeasures and decontamination procedures and technologies to protect their personnel on the battlefield. Due to its requirements, an internal security service can seek intelligence to detect, fully understand and intervene to prevent the theft of cultures of dangerous pathogens from a Government medical research facility which could be sold to adversaries (bioproliferation) or used in an attack (bioterrorism). The internal service could use this information to either mount intervention operations or support prosecutive actions against the perpetrators which it would be responsible for conducting.

5.3.3 *Some Aspects of Intelligence Missions, Authorities and Responsibilities Can Overlap Among and Between Agencies and Must Be Coordinated and Managed*

Plausible, Hypothetical Scenario: A professor who has recently "retired early" from a foreign university under seemingly odd circumstances has been determined to be actively seeking to consult on BSL-3/4 level infectious diseases to anyone who can pay his price. None of the companies and private investors he has contacted seem interested. However, he is contacted by a "consulting firm" (already under surveillance by the country's external intelligence service) who claims to be representing a "startup network" who will fund him with "unlimited resources and no regulatory controls to push genetic engineering of certain pathogens to the limit so that groundbreaking vaccines and therapeutics can be made to save mankind". The external service has assessed these claims as likely to be highly questionable. The professor mentions this new opportunity to a few close associates who are suspicious and contact authorities. The external and internal services agree to coordinate their activities to increase the likelihood of success and reduce conflicts and error. They both agree to share pertinent intelligence in a timely manner. The external service focuses on the full characterization of the "consulting firm", its members, the "startup network" and associations (including foreign countries or external groups), funding sources, the true nature of its activities. The internal service focuses on full extent of the professor's nature and activities with regard to so aggressively marketing himself, who he is willing to provide services to and who he might recruit or what he might access in-country to further his goals and objectives. Both are interested in how the professor, consulting firm and other parties would threaten their country's national security, and societal safety and resilience but from different perspectives; they both must be able to maintain operational security while gathering high value intelligence from their respective sources and methods in a timely manner and protecting sources and methods (neither can expose their own or each other's), including while converging on the same individuals, processes or locations.

5.3.4 The Nature and Types of Intelligence Collected, Processed, Analyzed and Used Can Vary Between Agencies

Plausible, Hypothetical Scenario: An unusual infectious disease outbreak has occurred in several towns along a disputed border region between Country AAA and Country BBB. The Country AAA Ministry of Health (MOH) deploys epidemiologists to investigate the outbreaks and public health specialists to treat those affected. The suspected agent and source are preliminarily determined. The MOH has encountered this organism in the past but not with the elevated number of infections or impacts. Unusual and heretofore unobserved properties of the causative organism are indicated. A nefarious attack cannot be ruled out at this juncture. The AAA National Police (NP) and Internal Security Service (ISS) are called in and initiate investigations and intelligence activities. This includes using microbial forensics capabilities they have at their disposal. (The AAA External Security Service (ESS) does not have such.) The samples that have been collected from patients, various scenes and specific matrices are split and analyzed in parallel according to both public health and microbial forensic protocols. The NP and ISS contact every laboratory that performs infectious disease research in Country AAA and eliminate them as possible sources of the etiologic agent. They also collect information from Country AAA experts about Country BBB experts who are knowledgeable about the organism in question. The ISS and NP request that the ESS focus intelligence activities on Country BBB experts and laboratories by all reasonable and prudent means. This request includes acquiring specimens of the organism of interest from specific laboratories that have been identified. If successful, forensic source attribution might occur and lead to identifying those involved so that national decision makers can direct appropriate actions to be taken. All operational and technical intelligence is shared between the three organizations which will use it for different, complementary purposes.

5.3.5 Complex Organization and Operational Constructs Exist Within and Between Agencies, That Must Be Navigated

Plausible, Hypothetical Scenario: A transnational, well-resourced, extremist terrorist organization has declared that they are pursuing and intend to use biological weapons against indigenous populations and foreign nations in their respective countries. They also claim on social media to have set up a "sophisticated laboratory that cannot be located or destroyed" to further their objectives. In one country that is concerned, HUMINT and SIGINT points to several specific locations from which the terrorist group is known to operate as possible laboratory sites. The country's intelligence community agrees that satellite imagery could be helpful. The control of the few IMINT satellites available is managed by one agency, and tasked by all of military services and the geospatial intelligence agency for their own priorities on particular schedules. This new priority is not on the current list so has to be inserted and addressed so as to be properly reposition the satellites but not to disrupt the current taskings of the other agencies. Senior officials and program managers from the various agencies the meet to study the new request and determine whether this new requirement, which is being levied by both their country's foreign intelligence and internal security services, can be satisfied. The request for SIGINT is also under review by the mission manager in the responsible agency.

5.3.6 Agencies Emphasize Different Methods and Uses of Collecting, Analyzing and Using Intelligence Based on Authorities, Missions, Assigned or Possessed Resources, Situations and Requirements

For expensive, resource- and infrastructure-intensive capabilities such as large complex SIGINT, GEOINT or IMINT, duplication of effort, investments and expenditures should be minimized for best value and return, provided requirements of customer agencies can be managed and addressed. It is unlikely that agencies that require occasional support from the above could effectively justify having their own capabilities of this magnitude. Even those that require support continually are better served by centralized management and resourcing than each having its own, except when specific capabilities are needed to address immediate and agency-specific requirements. Legal authorities, directives and interagency agreements may dictate where and how specific agencies can operate.

5.3.7 International and Interagency Engagements Are Tailored to Mission, Authorities, Needs and Resources

From my experience, international and interagency engagements by security services and law enforcement agencies are carefully managed and coordinated. With respect to the international, external services work with external services, internal services with internal services, and law enforcement with their counterparts. There are often legal, policy and security reasons for these arrangements. International engagements that require crossover to other domains are carefully adjudicated and managed to ensure that all parties are working within their authorities, equities are address and security is maintained. Within countries, the relationships and engagements can be much more fluid, particularly with respect to internal security and external security and internal security and internal law enforcement. One internal model that has worked well in the U.S. is the Joint Terrorism Task Force [14]. These exist within 104 major cities across the country and are comprised of Federal, State and local agencies. The JTTFs foster information and intelligence sharing, coordinated mission planning and response, more extensive leveraging of resources and expertise better tailored to requirements, and joint training and exercises.

5.3.8 Primary Agencies Can Have Relationships with Agencies or Others Which Can Provide Crucial Support

External and internal security agencies (and law enforcement organizations) may have technical support elements within which are used for day-to-day operations. However, it is very likely that specialized expertise and capabilities will be needed to inform and address specific needs and opportunities that arise. With the topic at hand, this could include deep expertise on specific pathogens and biotoxins, existing and emerging technologies and methods, or awareness of leading experts, research programs or laboratories. These resources exist and engage in government science and health agencies, academia, non-profit institutes and professional organizations.

By various mechanisms, security services (and law enforcement) can develop relationships which educate and inform in both directions and enable on-demand support when needed. To provoke thought by the reader as to how things work or could work in their own respective countries, I provide a few examples from the U.S.:

- The U.S. Federal Bureau of Investigation (FBI) routinely engages academic genetic engineering educational programs (e.g., IGEM, International Genetic Engineering Machine, 15) to educate students and faculty on biosecurity concerns and bioterrorism awareness; greater understanding of each parties' perspectives, expertise and equities are realized and relationships are developed which can be leveraged for other purposes. Other Federal agencies with pertinent interests are well aware of this initiative and coordinate with the FBI as appropriate.
- The U.S. Government national security and defense enterprise regularly funds the U.S. National Academies to organize open workshops and perform studies on priority topics. While these provide very useful and actionable insights and recommendations, they also help to build communities of experts that can be leveraged as needed for advice, consultation, training and specific studies. This strengthens national security, educates participants and breaks down barriers.
- When the U.S. stood up its microbial forensics investigative response program in the mid-1990s, among the first agencies that the FBI reached out to were the Centers for Disease Control and Prevention, and both the most prestigious Army and Navy infectious disease research programs. The FBI gained much needed access to expertise it did not have and initiated law enforcement understanding of biosurveillance and outbreak response which has resulted in robust coordination, the creation of the Laboratory Response Network (LRN, 16) which supports public health and law enforcement forensic investigative response, and information sharing that informs investigations and operations all the way to the upper reaches of the U.S. Federal law enforcement and national security community. Early on, the FBI also reached out to the Department of Agriculture for similar purposes though a similar laboratory network has not been formalized. In the U.S. the bulk of national biosurveillance capability is associated with the Departments of Health and Human Services (under which the Center for Disease Control and Prevention and Public Health Service exist) and Agriculture (under which pertinent agencies such as the Animal and Plant Health Inspection Service and Food Safety and Inspection Service reside). Public and agricultural health agencies are the most likely to first detect an usual or suspicious outbreak thus providing much-need indications and warnings to national security agencies. I am aware that relationships such as these exist in other countries, as well.

5.3.9 Within Agencies, Between Agencies with Shared or Complementary Missions, Capabilities and Infrastructure, and Across Intelligence Enterprises, Coordination and Deconfliction Are Critical

There are often mission-specific reasons for which coordination and deconfliction can be very advantageous, but these phenomena also extend more broadly to:

- Enterprise planning, integration and prioritization
- Major infrastructure and capability investments
- Budget formulation and execution
- Reducing duplication of effort and investments, amplifying or extending return on investments where possible
- Managing and adjudicating collection and analysis requirements which involve shared capabilities or resources
- Managing and protecting sensitive operations, sources and methods
- Anticipating and managing risks and failures
- Managing external resources that are being tasked

A formalized interagency, including for specific areas of interest such as intelligence collection, analysis and use for bioterrorism prevention, intervention, and control normalizes and promotes additional benefits which transcend myriad stochastic and ad hoc relationships and transactions to:

- Delineate responsibilities while identifying shared and complementary missions, goals, objectives and outcomes sought
- Identify and formalize relationships and mechanisms to leverage specialized resources and assets in other agencies to minimize duplication of effort and maximize efficiency
- Share information and intelligence that can be used in ways other than the original, including situational awareness
- Share value and return on investment for research, development, test and transition of new technologies and capabilities
- Increase tradecraft, operational, analytical and technological innovation opportunities
- Extend trusted networks that can be leveraged for mission, operation and project-specific requirements
- Establish equitable deconfliction and adjudication mechanisms
- Optimize innovation and integration processes

5.3.10 *Intelligence Informs Decision Making and Ultimately Outcomes at the Tactical, Operational and Strategic Levels*

Ultimately, the purpose of intelligence is to inform decisions and courses of action. When this occurs, certain conditions or phenomena should be taken into consideration which will influence how requirements are addressed, capabilities are used, processes function, risks and uncertainties are managed and outcomes can be projected and used:

- Fundamentally, intelligence exists and is pursued to increase knowledge, understanding, ideally prospectively but also concurrently and retrospectively
- Decisions requiring intelligence support are often complex, and are made on a timeline; these timelines can be fast moving and rapidly changing
- Intelligence is a dynamic, adaptive process; sometimes decision making can be but sometimes not
- Uncertainty and risk has to be accounted for and communicated with intelligence to decision makers to inform the best decisions possible
- Intelligence is rarely "perfect"
- Intelligence professionals routinely seek to anticipate and manage risks or exigencies, however suboptimal outcomes or failures do occur
- Intelligence supports users at the tactical, operational and strategic levels; needs can overlap but often differ, i.e., level of fidelity, timeliness and responsiveness of delivery
- Senior officials (users) are notoriously impatient and demanding with intelligence providers

If a nation is faced with a decision of whether to create or improve its intelligence capabilities against bioterrorism, bioproliferation and biowarfare, these options and hypothetical outcomes might help focus discussions. Imagine that an adversary which has sufficient operational, financial, technical and logistical resources to execute an attack which would result in substantial negative impact. What might the outcomes be if the targeted nation had no, limited and extensive capabilities?

- No Intelligence:
  - Planning, preparations and attack occur, objectives of adversary achieved, impact is considerable and evolving
  - Adversary is emboldened, considers other attacks with other outcomes; takes credit publicly and issues threats on social media including claiming credit for natural outbreaks as attacks; with no apparent action by the government, the news media increases their calls for action and questions its competency
  - Targeted country population sustains psychosocial effects, to include loss of confidence in Government to protect them; fear escalates; societal patterns are altered resulting in changes to daily life; instability in Government becomes noticeable

- Partial Intelligence:
  - The Government has most likely persons/groups under surveillance as it becomes aware; Because resources are limited, the Government purposely exposes some of its activities to some targets to disrupt their planning and activities; Technical reachback is available to identify threat agents; informants being developed but not in place
  - Adversary believes that all communications and activities are being monitored, erratically changes methods to "achieve security"; psychological effects of Government operations are having effects on some suspected group members and enablers
  - Attack is conducted, but more limited in scope than originally planned and does not achieve most initial objectives
  - Two persons of interest are immediately apprehended and interrogated; physical evidence (including samples of threat agents used and dissemination devices) is collected and is being exploited, including for attribution purposes
- Perfect Intelligence
  - As soon as potential adversary begins to organize a cell and initiate planning, Government intelligence assets are trained on communications, funding, physical movements, acquisitions and logistics, bases of operation and recruitment activities
  - Government allows adversary to plan and prepare for attack, but surreptitiously intervenes through a third party and substitutes a benign threat agent for the harmful one the adversary sought to acquire
  - As soon as the adversary actually begins the operation and mobilizes, the Government intervenes and disrupts, apprehends all principals and initiates interrogations, seizures of properties, equipment and materiel, and all adversary computers, documents and communications devices. Cooperation with the Ministry of Justice is initiated for prosecution while intelligence operations continue on the adversary's network

From my perspective, one should also integrate the **Realities of Intelligence** into one's calculus to include:

- Effective and timely intelligence which anticipates, responds to and predicts against a diverse array of threats, known and unknown, is very difficult and often complex—a lot easier said than done
- A comprehensive intelligence system is expensive to establish and maintain
- Intelligence by its nature is fraught with gaps and uncertainties that must be understood and accommodated, at all levels and applications—as once described, it is a "wilderness of mirrors"
- One intelligence method, tactic or technique is rarely sufficient to meet a particular requirement (need) or inform a particular desired outcome. A choreography of methods and techniques, often delicate, is often required

- Methods, sources, tactics, techniques, procedures and resources must be available, properly mature and positioned or can be called up and tailored on demand; it often takes considerable time, risk management, ingenuity and innovation, and expense for the "tool kit" to exist and be available
- Intelligence is a government domain, by necessity it must be protected from adversaries and usually is kept from unwarranted public awareness and view. Otherwise, it becomes weakened or useless. Even this requires substantial planning, care and investments

5.4 Key Considerations for Intelligence Design, Preparation and Execution

Unless one understands what one could be facing or expected to encounter, designing, positioning and using capabilities could go very wrong. Thus, from the bioterrorism prevention and control perspective and this treatment of intelligence in this context, these elements should be considered:

At a high level, one might ask what situations could or would be expected, e.g. involving or related to:

- A suspected bioterrorist, facilitator or bioproliferator
- A facility in which illicit activities are suspected
- An event of interest, including suspicious outbreaks
- Possible threats which could result from the misuse of knowledge and technology ("dual use")
- At a bit higher fidelity, plausible situations could include:
  - High Impact Bio Attack (Small to Large), No Warning
  - High Impact Attack (Small to Large), Warning
  - Surgical Strike, Military vs. Civilian Targets
  - Suspicious or Anomalous Disease Outbreak
  - Intelligence, Investigation Suggesting Imminent Attack
  - Attack Which Targets a Special High-Level Event
  - "Stumble Upon" a Cache, Suspect Laboratory or Storage Facility
  - Indications and warnings Regarding Suspicious or Illicit Activities in a Laboratory Facility
  - Possible Attack or Claim Which Turns out to be a Hoax

Note: Each one of the above could be developed into a set of likely planning scenarios that could aid capability development, experimentation and use.

It is also crucial to understanding an adversary or "person of interest" to the greatest degree possible:

- What are adversary goals and objectives?
  - Inflict terror and fear
  - Influence and even control behaviors of targets

- Power projection (against a government, society, populations)
- Territorial control
- Cause harm (death, persistent negative outcome, destruction)
- Convey political or other message
- Replace or offset political system
- Cause authority to expend considerable resources to defend or protect itself
- Increase own visibility, influence and power with respect to competitors
- Financial gain
- Revenge

Note: Do any, some or all of the above influence planning, investment choices, implementation and how intelligence capabilities are to be used?

In order to accomplish their goals and objectives, adversaries choose targets that could include:

- Specific Populations
- Consumables
- Commodities
- Infrastructure Systems
- High Value Facilities and Venues
- Special Events
- National Symbols
- Mass Psychology and Perception
- Social, Societal and Political "Fabric"

Notes: Do intelligence capabilities have to address what targets could be chosen and the means, methods and timing of an adversary's planning, preparation and attack? Would variation be required to address each or all of them satisfactorily?

Like most, an adversary has a "business cycle" which includes major steps, such as:

- Developing Concepts & Intentions
- Designing Programs and Envisioning Outcomes
- Preparing Plans for Targeting, Weapons, Tactics, Techniques and Procedures
- Engaging in Acquisition, Logistics and Movement
- Experimentation with Weapons and Delivery Methods
- Production & Testing of Weapons and Delivery Method
- Training, Preparation and Rehearsal
- Conducting Operations & Attack
- Conducting Evasion & Escape
- Claiming Responsibility or Engaging in Denial & Deception

Note: How would intelligence capabilities be best designed, tested, configured and used against any or all activities in each to inform decisions and actions? How much and what kind of information is needed?

Other considerations from the adversary point of view that could be integrated into how one plans, develops, stages and applies specific intelligence capabilities could include:

- Resources, Equipment and Materiel Required
- Recruitment of Personnel and Specific Expertise
- How, When and Where Planning and Preparation Occurs
- Research, Development, Test and Transition, Sites, Approaches and Tolerances
- Operational Environment Selected and Security Applied
- Personnel, Operational and Communications Security
- Financial Sources, Support and Network, Security Methods
- Supply Chain: Procurement or Acquisition and Logistics, Security Approaches
- Staging and Operational Bases, Security
- Transport and Movement of Personnel, Equipment, for Staging and Operations

Note: Each one of these is a sub-system that, in itself, does require detailed understanding to most effectively address a threat, i.e., adversary, methods, means, modes and outcomes sought.

Lastly, it is crucial to have a deep technical understanding of what weapon an adversary might choose, how it could be weaponized, delivered, and what the intended vs. actual effects would be. For this, one can use various sources of information such as biothreat lists, such as published by responsible Government agencies or programs [17, 18] and interactions with experts who understand the biology, ecology, public health impact and weaponization and dissemination of pathogens and biotoxins, to begin to understand what one might be dealing with. (For brevity's sake, agricultural pathogens are not included.) If a nation's national security community or intelligence community does not have expertise in these organisms, or the foundational science and technology (such as rapidly emerging genetic engineering methods, genomic sequencing and exploitation, gene synthesis and synthetic biology), they should acquire it. Unless they expect to hire such, they should go to trusted expert sources in those agencies and vetted expert sources within external institutions (Table 5.1).

Alternatively, constructs can be used which are designed to "operationalize" how existing or emerging threats might be leveraged or be employed, and how existing or planned intelligence capabilities, particularly involving technical assets, would have to be optimized to be most effective, for example:

- "New strain" of influenza (highly aggressive), covert attack during flu season
- Foot and Mouth Disease (FMD) clandestinely introduced in large animal feed lot which becomes distributed in system
- Large scale anthrax attack
- Aerosolized ricin disseminated in closed or contained environment
- Botulinum toxin clandestinely and selectively introduced in food supply, small scale, distributed attacks
- Any biological weapon, for which plausible denial and deception has been engineered in weapon and modus operandi

Table 5.1 Bioterrorism threat agents, U.S. Centers for Disease Control and Prevention (CDC) [17]

Category A:
Can be easily disseminated or transmitted from person-to-person; results in high mortality and have the potential for major public health impact; might cause public panic/social disruption; requires special attention for public health preparedness
Anthrax, Botulinum toxin, Plague, Smallpox, Tularemia, Viral Hemorrhagic Fevers (e.g., Ebola, Marburg)
Category B:
Are moderately easy to disseminate, result in moderate morbidity and low mortality rates, require specific enhancements of CDC diagnostic capacity and enhanced disease surveillance
Brucellosis, Epsilon toxin of Clostridium perfringens, Food safety threats (e.g., Salmonella, Escherichia coli O157:H7, Shigella), Glanders, Melioidosis, Psittacosis, Q Fever, Ricin toxin, Staphylococcus enterotoxin B, Typhus fever, Viral Encephalitis (Venezuelan Equine Encephalitis, Eastern Equine Encephalitis, Western Equine Encephalitis), Water safety threats (e.g., Vibrio cholerae, Cryptosporidium parvum)
Category C:
Future threats which could be engineered for greater availability, ease of production and dissemination, potential for high morbidity and mortality
Also, Nipah and Hanta viruses

- Effective attack of any scale using as yet unknown, "naturally occurring" strain or genetically engineered pathogen with enhanced disease or antimicrobial or antiviral properties
- Assassination of unprotected VIP using biotoxin, considerable operational security or serendipity by perpetrator (planning through execution)

One might also consider integrating the following points into study, discussion, decisions, planning and choices that are made for intelligence capabilities and support thereof:

- What is needed to achieve sufficient and timely understanding of adversaries that have expressed or suggested interest in biological weapons, and their enablers?
- What situational awareness is required regarding persons and facilities performing or having legitimate access to biological threats agents and key equipment?
- What detailed knowledge exists and must be available on demand for potential biological threat agents and sources?
- What knowledge exists or has to be developed for a thorough understanding of our own or others' vulnerabilities and how should these be prioritized for the purpose at hand?
- What "Open Source" biosurveillance information on infectious disease outbreaks and patterns exists and what is needed to inform early indications and warnings of suspicious outbreaks?
- What knowledge and expertise is available to provide detailed awareness of current and emerging knowledge and technology and their "dual use" properties?
- What deployable technical support of various methods and means is available or needed for detection, characterization and attribution purposes?

To wrap up, on top of all the challenges and complexities associated with intelligence writ large, intelligence focused on the prevention and control of biological weapons acquisition, development and use has additional considerations to contend with, which are different from other Weapons of Mass Destruction threats and concerns:

- There are myriad threats, targets, situations, scenarios and variations of these
- Bioterrorism and bioproliferation activities can easily be masked or obscured from view or notice, or even be "hid in the open" (e.g., "dual use")
- Knowledge, source materials and technology is readily available, including source materials from nature
- Potential perpetrators (and enablers) are considerably numerous
- Activities of concern are scalable
- Neither sophisticated or large-sized operations are necessary to be successful
- Specialized knowledge and capabilities are needed to address intelligence, operational, technical, policy and legal requirements and needs that are significantly different than other WMD
- Decision makers, advisors, stakeholders and customers may understand the nuclear or chemical threat but do not necessarily understand the biological threat; thus, resources should be created or made available to assist with their understanding and decision making

5.5 Managing the Time-Risk Continuum

With all of the complexity, uncertainty and challenges which fold into a strategic system of intelligence which includes countering bioterrorism, bioproliferation and biowarfare, the "bottom line" is that nations are faced with managing risk in many dimensions and contexts, as have been discussed thus far. But, there is one more. The time component also influences how capabilities are designed, configured, implemented and used. Ideally, these should effectively deal with exigent, unexpected circumstances as well as standing requirements. The Time-Risk Continuum concept is depicted in Fig. 5.5.

Every process, action or event has a timeline associated with it. The timeline can be short, compressed and streamlined or it can be prolonged, more extensively developed and complex. To achieve intelligence objectives, not only do capabilities have to be applied *where* and *how* they will provide the most benefit, they also must be applied *when* they achieve such.

The author posits that the overall goal is to move as far to the left as possible. Other than with the unlikely situation of chance perpetually being in one's favor, the only effective way to "drive to the left" is with sufficiently timely and complete foreknowledge.

This goal is enabled and achieved by:

- Possessing established, positioned, resourced and replenished Intelligence Assets
- Having laws and policies in place that support or allow use

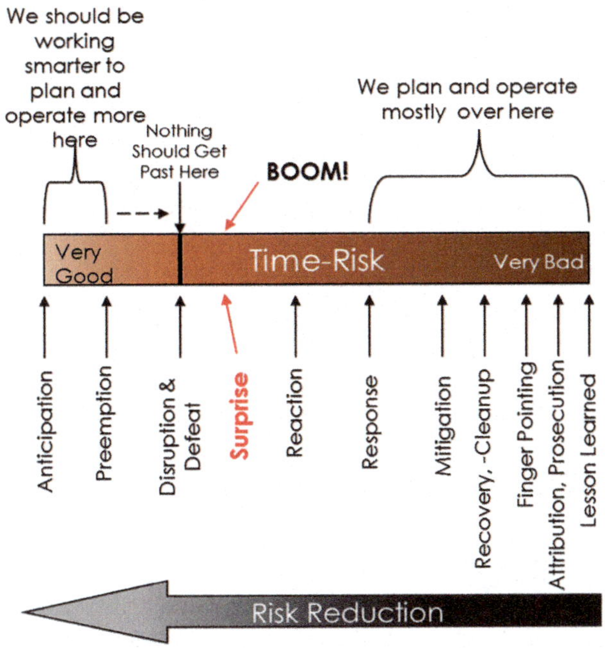

Fig. 5.5 The time-risk continuum concept for reducing security risk

- Having the ability to tailor and apply intelligence collection and analysis assets as needed
- Being capable of applying tailored technical support and expertise as needed
- On-demand reachback capabilities which are prepared and staged
- Timely and actionable intelligence and information sharing to support operations and analysis
- Maximal utility and credibility of intelligence and other key information to support decision making

5.6 If One Has Nothing or Little, How Does One Get Started?

The author argues for the importance of a vision, strategy and plan that articulates goals, objectives, priorities, desired outcomes and perhaps measures of effectiveness. Documents of this sort are most useful whether one is just starting or to improve the enterprise.

Such a document or collection of documents:

- Communicates a common purpose and unified vision for the desired state
- Sets expectations for participants and contributions by each

- Assigns respective roles, authorities and responsibilities
- Develops and communicates priorities
- Provides goals and objectives and general guidance to achieve
- Establishes coordination, collaboration and leveraging relationships and functions
- Depicts a trajectory from current to desired state
- Provides an "architecture" (overall design) for the enterprise
- Leads to plans, concepts of operations
- Positions an agency or interagency to measure effectiveness

The interagency engagement for strategy or plan development can provide a fruitful and contemporaneous systems analysis of existing capabilities, how they can be tailored for application against the instant set of threats, and the identification and prioritization of gaps. All of this would be fed into the strategy, plan, program and capability development, budgeting and interagency coordination. In order for success to be achieved, one organizational component has to take the lead and continuity of commitment has to occur.

Even if one does have some capabilities and they are spread across various organizations, a strategic vision or plan has considerable value. It conveys importance, unifies and aligns agencies and their respective capabilities, ensures that priorities are stated and included in planning and budgeting and can be used to measure progress with the proper metrics.

5.7 Concluding Points, a Possible Way Forward

In summary, I have the unshakable view that intelligence is critical to prevent, disrupt, mitigate, respond to and attribute bioterrorism, bioproliferation or biowarfare, even though additional capabilities and resources are also needed. For nations that do not have or have minimal intelligence capabilities for these endeavors, I suggest that this chapter can provide a basis for assessing whether it is needed, what capabilities and considerations should be developed, acquired and tailored against priorities, requirements and gaps, and how to organize and sustain for best value and outcomes.

I realize that I have provided a great deal of information for consideration. I have not presumed to know what all readers would want or would act upon should they be in a position to or interested in doing so. Rather, the choices are left to the reader for their own interests and purposes.

Should a nation or alliance decide to undertake a national assessment or planning activity, it might consider seeking assistance. Certainly, professionals within a government aided by in-country experts would have to participate and take ownership. However, such a process could be facilitated with likelihood of success increased by engaging external expertise.

Given the magnitude and complexity of such an undertaking, I suggest that NATO could be a crucial resource. A couple of suggestions of ways through which the Alliance could lead in this area are through conferences or workshops, or even a series of carefully crafted events. One workshop could provide those countries that would like to create or enhance these capabilities the opportunity to do so, through a structured process led by others that have established their own. A second workshop could provide alliance partners, experienced and inexperienced, with the opportunity to leverage universities and "think tanks" for new ideas and innovations in this arena.

Further, NATO countries which have mature enterprises could advise those which do not how these can be best organized and function. Assuming equities can be properly managed, those countries which have well-developed capabilities might be called upon to provide continuing consultation for those which do not. Certain countries within the Alliance have existing technical response and forensic investigation programs could provide advice and technical support to those which do not. The relationship between government agencies and external experts and technical resources could be the foundation for workshops or other formats to bring these elements together, establish relationships and focus on achievable objectives. Bringing emerging young experts into the discussion provides a "pipeline" for the future.

Whichever is chosen and pursued, success begins with vision and leadership.

References

1. Sun Tzu (trans: Lionel Giles). The art of war. ISBN-13:978-1545211953
2. Bimfort MT (1995) A definition of intelligence. Stud Arch Ind 2(4). Center for the Study of Intelligence, Central Intelligence Agency. https://www.cia.gov/library/center-for-the-study-of-intelligence/kent-csi/vol2no4/html/v02i4a08p_0001.htm
3. Van Cleve M XXXX Strategic counterintelligence. Stud Intell 51(2). Center for the Study of Intelligence, Central Intelligence Agency. https://www.cia.gov/library/center-for-the-study-of-intelligence/csi-publications/csi-studies/studies/vol51no2/strategic-counterintelligence.html
4. National Security Act of 1947 Public Law 253, 80th Congress (Last Amended 1966)
5. Johnson LK (ed) (2007) Handbook of intelligence studies. Routledge, Abingdon/Oxon/New York, 381 pp
6. Johnson LK (2010) Oxford handbook of national security intelligence. Oxford University Press, New York, 886 pp
7. Wikipedia (2017) List of intelligence gathering disciplines. https://en.wikipedia.org/wiki/List_of_intelligence_gathering_disciplines
8. National Intelligence Council (2017) Global trends: paradox of progress. National Intelligence Council, Washington, DC. https://www.dni.gov/files/documents/nic/GT-Full-Report.pdf
9. National Intelligence Council (2016) Implications for US National Security of anticipated climate change. NIC WP, Washington, DC.2016-01
10. Clark RM (2016) The protection of intelligence sources and methods. In: Oleson PC (ed) AFIO's guide to the study of intelligence. The Association of Former Intelligence Officers, Falls Church, pp 587–596

11. Federal Bureau of Investigation. https://www.fbi.gov/about/leadership-and-structure/intelligence-branch
12. New Zealand Security Intelligence Service. http://www.nzsis.govt.nz/our-work/our-methods/
13. Collier's International (2012) Spotlisght: U.S. Intelligence Community. http://www.capitolmarkets.com/spotlight/spotlight-the-us-intelligence-community/
14. Federal Bureau of Investigation, Joint Terrorism Task Forces. https://www.fbi.gov/investigate/terrorism/joint-terrorism-task-forces
15. International Genetic Engineering Machine (IGEM). http://igem.org/Main_Page
16. Laboratory Response Network, Centers for Disease Control and Prevention. (2014) https://emergency.cdc.gov/lrn/
17. Centers for Disease Control and Prevention (2017) Bioterrorism Agents/Diseases. https://emergency.cdc.gov/agent/agentlist.asp
18. Federal Select Agent Program (2017) www.selectagents.gov/Select%20 Agents%20and%20Toxins%20List.html

Chapter 6
Comparison of the Available Methods of Differentiation Between a Biological Attack and Other Epidemics

Goran Belojevic

Abstract Timely detection of a deliberate epidemic and the engagement of security and law enforcement forces are of crucial importance, because breaking the epidemiological chain does not guarantee that evil human minds will not start another epidemic. There are four available methods of differentiation between a biological attack and other epidemics: Grunow and Finke (Clin Microbiol Infect 8:510–521, 2002), Dembek et al. (Epidemiol Infect 135: 353–371, 2007), Radosavljevic and Belojevic (Public Health 126: 77–81, 2012) and Radosavljevic (Biopreparedness and public health. Springer, Heidelberg, pp 17–32, 2013). The aim of this study is to compare the application of these four methods on three documented bioterrorist attacks (Salmonellosis – The Dalles, Oregon, 1984; Shigellosis – Dallas, Texas, 1996; Anthrax USA – 2001) one accidental release of a weaponized agent (Anthrax – Sverdlovsk, Soviet Union, 1977) and three unusual epidemiological events (West Nile Virus – NYC 1999; Tularemia, Kosovo, 2000; Escherichia coli O104-H4 outbreak in Germany 2011). The results show that four methods are closely related in differentiation between a biological attack and other epidemics. Dembek et al. method and Radosavljevic & Belojevic method are simplier and most effective during an epidemic. Grunow & Finke method is more complex and most effective after an epidemic. Radosavljevic method is most detailed and allows for a further differential analysis of an unusual epidemiological event.

G. Belojevic (✉)
Institute of Hygiene and Medical Ecology, Faculty of Medicine, University of Belgrade, Belgrade, Serbia
e-mail: goran.belojevic@med.bg.ac.rs

V. Radosavljevic et al. (eds.), *Defence Against Bioterrorism*, NATO Science for Peace and Security Series A: Chemistry and Biology,
https://doi.org/10.1007/978-94-024-1263-5_6

6.1 Introduction

Early detection of a bioterrorist attack is crucial in order to engage security and law enforcement forces timely, because differing from natural epidemics, breaking an epidemiological chain is not a sufficient measure against bioterrorism [12, 14] . Evil human mind might be willing to start another epidemic and perpetrators have to be neutralized as soon as possible. There are four available epidemiological methods in literature for differentiation between a biological attacks and other epidemics [4, 6, 13, 15].

It is unclear how the four differentiating methods correlate when applied on specific unusual epidemiologic events (UEE). The aim of this study is to compare the application of these four methods on various UEEs including documented cases of bioterrorism, biological accidents and outbreaks that were suspected to be intentionally caused.

6.2 Material and Methods

There are only four available methods of differentiation between biological attack and other epidemics [4, 6, 13, 15] which we apply on several UEEs: three documented bioterrorist attacks (Salmonellosis – The Dalles, Oregon, 1984; Shigellosis – Dallas, Texas, 1996; Anthrax USA – 2001), one accidental release of a weaponized agent (Anthrax – Sverdlovsk, Soviet Union, 1977) and three unusual epidemics (West Nile Virus – NYC 1999; Tularemia, Kosovo, 2000; Escherichia coli O104-H4 outbreak in Germany 2011). This is a post hoc analysis with a time distance that allows for input of some data that were not known at the time of an epidemic.

Using the Grunow and Finke method [6], an UEE is assessed qualitatively with 11 non-conclusive criteria of indirect importance and two conclusive criteria which directly prove that a biological attack has taken place (identification of the agent as a warfare agent and proof of the release of the agent as a biological weapon). Each non-conclusive criterion is assessed and scored from 0–3 based on the obtained data – 0 (no available data or the rejected criterion); 1 (there is a suspicion of the use of a biological weapon but with a considerable level of uncertainty); 2 (there are still unclear indications of the use of a biological weapon); 3 (there are clear indications or proofs of a biological attack). Each non-conclusive factor is also weighted from 1–3 based on its usefulness for detecting a biological attack – 1 (natural causes can explain the criterion); 2 (there is only a limited possibility to explain the criterion by natural casues); 3 (the criterion cannot mainly or fully be explained by natural causes). Assessment points for each non-conclusive factor are multiplied by a weighting factor to obtain a total number of points which allows reasoning about the likelihood of a biological attack – 0–17 (unlikely); 18–35 (doubtful); 36–50 (likely) and 51–54 (highly likely).

The Dembek et al. method [4] uses 11 potential epidemiological clues to discern between deliberate and natural infectious diseases outbreaks.

With the Radosavljevic and Belojevic method [15], an UEE is assessed using 10 indicators each scored with 0 or 1 if there is a low or high probability of a deliberate or accidental outbreak, respectively. Reasoning about the possibility of a deliberate or accidental outbreak is based on the total score – 1–4 (probably natural outbreak); 5–7 (possibly deliberate or accidental outbreak); 8–10 (probably deliberate or accidental outbreak).

The Radosavljevic method [13] allows for a detailed differentiation between a biological attack, a natural epidemic, an outbreak of a new or re-emerging disease or an accidental release of a pathogen. The method is based on 33 indicators, 23 qualitative and 10 quantitative. These indicators are related to four components of an UEE: 1. Perpetrator/source/reservoir of infection; 2. Agent/pathogen; 3. Means/media of delivery/transmission mechanisms; and 4. Target/susceptible population. Each indicator is scored either with N/A (non-applicable) or with 0 and 1 for low and high probability of a specific UEE, respectively. If any of the four components is scored only with 0 or N/A the related scenario is eliminated from further consideration. Total scores for each outbreak scenario may range from 0–33 and the probability of a specific outbreak is assessed as follows: 0–8 = Lowly probable; 9–16 = Possible; 17–24 = Highly probable; 25–33 = Certain.

6.3 Results

6.3.1 Salmonellosis – The Dalles, Oregon, 1984

There were two large cohorts of Salmonellosis cases in Dalles, Oregon, from 9 September to 10 October 1984. There were 751 cases, and 45 of them were hospitalized. The common feature for the cases was dining in some of 10 restaurants with salad bars. *Salmonella Typhimurium* was isolated from 388 patients and from food items. In October 1985 a vial with the same strain as the outbreak strain was found by the FBI in the nearby Rajneshee cult clinic laboratory. It was purchased prior to the outbreak. Two cult members were arrested and served federal prison terms [18].

Using the Grunow and Finke method in assesing this outbreak a total score of 22 indicates that the use of a biological warfare agent is doubtful (Table 6.1).

When the Dembek et al. method is applied on this outbreak four out of 11 clues are positive, but there is a direct evidence, indicating a high probability of a biological attack (Table 6.2).

Using the Radosavljevic & Belojevic method this UEE is assessed as a possibly artificial outbreak as a total score is 5 (Table 6.3).

If the Radosavljevic method is used to assess this UEE total score of 12 indicates that this outbreak is a possible biological attack. (Table 6.4).

Table 6.1 Assessment of the Salmonellosis outbreak – the Dalles, Oregon, 1984 with the Grunow & Finke method (2002)

No.	Criterion	Assessment (Points)	Weighting factor	Points
1.	Existence of a biological risk	3	2	6
2.	Existence of a biological threat	0	3	0
3.	Special aspects of a biological agent	1	3	3
4.	Peculiarities of the geographic distribution of the biological agent	2	1	2
5.	High concentration of the biological agent in the environment	0	2	0
6.	Peculiarities of the transmission mode of the biological agent	3	1	3
7.	Peculiarities of the intensity and dynamics of the epidemic	2	2	4
8.	Peculiarities of the time of the epidemic	1	1	1
9.	Unusually rapid spread of the epidemic	3	1	3
10.	Limitation of the epidemic to a specific population	0	1	0
11.	Peculiarities of the clinical manifestation	0	1	0
	Total			22

Table 6.2 Assessment of the Salmonellosis outbreak – the Dalles, Oregon, 1984 with the Dembek et al. method (2007)

No.	Clue	Assessment
1.	A highly unusual event with large numbers of casualties	Positive
2.	Higher morbidity or mortality than expected	Negative
3.	Uncommon disease	Negative
4.	Point source outbreak	Positive
5.	Multiple epidemics	Positive
6.	Lower attack rates in protected individuals	Negative
7.	Dead animals	Negative
8.	Reverse spread	Negative
9.	Unusual disease manifestation	Negative
10.	Downwind plume pattern	Negative
11.	Direct evidence	Positive
	Total	4

Table 6.3 Assessment of the Salmonellosis outbreak – the Dalles, Oregon, 1984 with the Radosavljevic & Belojevic method (2011)

No.	Epidemiological indicator	Score (1 = Yes; 0 = No)
1.	Unusual/atypical manifestation of a known disease	0
2.	Several unusual/unexplained syndromes coexisting in the same case without any other explanation	0
3.	A sudden unexplainable increase in the number of cases or deaths in human populations	1
4.	Higher than expected morbidity and or mortality rates	1
5.	Clustering of patients with fever only or with fever and other symptoms	1
6.	A disease identified in the region for the first time again after a long period of time or after its eradication	1
7.	A new strain of pathogen identified in the region for the first time after a long period or after its eradication	0
8.	A disease with an unusual/atypical seasonal distribution	0
9.	One or more explosive epidemics/outbreaks with indicators of point-source origin	1
10.	A disease with an unusual geographic distribution.	0
	Total	5

Table 6.4 Assessment of the Salmonellosis outbreak – the Dalles, Oregon, 1984 with the Radosavljevic method (2013) (1 = Yes; 0 = No)

No.	Perpetrator/Source of epidemic/Reservoir of pathogen	Biological agent/Pathogen	Means/Media/ Factors of transmission	Target/ Susceptible population
1.	Sophistication (0)	A category (CDC) (0)	Air (0)	Intelligence (0)
2.	Motivation (1)	B category (CDC) (1)	Food (1)	Secrecy (0)
3.	Intention (1)	C category (CDC) (0)	Water (0)	Personal control (0)
4.	Intelligence (0)	Emerging pathogen (0)	Fomites (0)	Control of means of delivery (0)
5.	Secrecy (1)	Amount of the pathogen (1)	Vectors (0)	Physical protection (0)
6.	Number of perpetrators (1)		Biological ammunition (0)	Protection by chemoprophylaxis (0)
7.	Number of sources of infection/reservoirs (1)		Delivery systems (0)	Protection by immuno-prophylaxis (0)
8.	Accessibility to sources of agent/pathogen (1)		Dispersion systems (0)	Importance of a target/ population (0)
9.	Accessibility to targets/ population at risk (1)			Location of a target/ population (1)
10.				Size of a target population (1)
11.				Distribution of a target population (0)
	Total score 12			

6.3.2 *Shigellosis, Dallas, Texas, 1996*

The outbreak of acute diarrhoeal disease among 12 laboratory workers from St. Paul Medical center in Dallas lasted from 29 October to 1 November 1996. All of them ate doughnuts and muffins left in their break room on 29 October. *Shigella dysenteriae* type 2 was found in the stools of eight workers. An anonymous e-mail sent from the supervisor's computer invited workers to eat pastries in the laboratory break room on 29 October. The mean incubation period was 25 h. Investigation suggested that the source of the outbreak was S. dysentreiae type 2 isolates kept in the hospital laboratory storage freezer. On 28 August 1997 a laboratory technician was convicted and sentenced to 20 years in prison [9].

With the Grunow and Finke method, it is assessed that the use of a biological warfare agent in this UEE is doubtful as a total score is 19 (Table 6.5).

If the Dembek et al. method is used to assess this UEE, three out of 11 clues are positive, but there is a direct evidence, indicating a probable biological attack (Table 6.6).

When the Radosavljevic & Belojevic method is used to assess this UEE, a total score of 5 indicates a possibly artificial outbreak (Table 6.7).

Using the Radosavljevic method in assessing this UEE a total score of 12 indicates a possible biological attack. (Table 6.8).

Table 6.5 Assessment of the Shigellosis outbreak – Dallas, Texas, 1996 with the Grunow & Finke method (2002)

No.	Criterion	Assessment (Points)	Weighting factor	Points
1.	Existence of a biological risk	0	2	0
2.	Existence of a biological threat	0	3	0
3.	Special aspects of a biological agent	2	3	6
4.	Peculiarities of the geographic distribution of the biological agent	2	1	2
5.	High concentration of the biological agent in the environment	0	2	0
6.	Peculiarities of the transmission mode of the biological agent	3	1	3
7.	Peculiarities of the intensity and dynamics of the epidemic	0	2	0
8.	Peculiarities of the time of the epidemic	1	1	1
9.	Unusually rapid spread of the epidemic	3	1	3
10.	Limitation of the epidemic to a specific population	3	1	3
11.	Peculiarities of the clinical manifestation	1	1	1
	Total			19

Table 6.6 Assessment of the Shigellosis outbreak – Dallas, Texas, 1996 with the Dembek et al. method (2007)/positive clues only/

No.	Clue
3.	Uncommon disease
4.	Point source outbreak
11.	Direct evidence

Table 6.7 Assessment of the Shigellosis outbreak – Dallas, Texas, 1996 with the Radosavljevic & Belojevic method (2011)/indicators scored with 1/

No.	Epidemiological indicator
3.	A sudden unexplainable increase in the number of cases or deaths in human populations
4.	Higher than expected morbidity and or mortality rates
5.	Clustering of patients with fever only or with fever and other symptoms
6.	A disease identified in the region for the first time again after a long period of time or after its eradication
9.	One or more explosive epidemics/outbreaks with indicators of point-source origin

Table 6.8 Assessment of the Shigellosis outbreak – Dallas, Texas, 1996 with the Radosavljevic method (2013) (indicators scored with 1/

No.	Perpetrator/Source of epidemic/Reservoir of pathogen	Biological agent/Pathogen	Means/Media/Factors of transmission	Target/ Susceptible population
1.	Sophistication	B category (CDC)	Food	Location of a target/population
2.	Motivation	Amount of the pathogen		
3.	Intention			
4.	Intelligence			
5.	Secrecy			
6.	Number of sources of infection/reservoirs			
7.	Accessibility to sources of agent/pathogen			
8.	Accessibility to targets/ population at risk			

6.3.3 Anthrax –USA 2001

Anthrax-containing letters were placed in the mail to several news media offices and two senators. There were 22 cases of anthrax (11 inhalational and 11 cutaneous) and five of them died. The mail-related isolates were indistinguishible and came from the same source. The strain was traced to the military microbiological laboratory at Fort Detrick [7, 8].

If the Grunow and Finke method is used to assess this UEE, a total score of 51 indicates that intentional use of a biological warfare agent was highly likely (Table 6.9).

By the Dembek et al. method in assessing this UEE, four out of 11 clues are positive, but there is a direct evidence indicating a probable biological attack (Table 6.10).

When the Radosavljevic & Belojevic method is used to assess this UEE, a total score of 9 indicates a probably artificial outbreak (Table 6.11).

Using the Radosavljevic method in assessing this UEE, a total score of 17 indicates a highly probable biological attack (Table 6.12).

Table 6.9 Assessment of the Anthrax outbreak – USA 2001 with the Grunow & Finke method (2002)

No.	Criterion	Assessment (Points)	Weighting factor	Tot
1.	Existence of a biological risk	3	2	6
2.	Existence of a biological threat	2	3	6
3.	Special aspects of a biological agent	3	3	9
4.	Peculiarities of the geographic distribution of the biological agent	3	1	3
5.	High concentration of the biological agent in the environment	3	2	6
6.	Peculiarities of the transmission mode of the biological agent	3	1	3
7.	Peculiarities of the intensity and dynamics of the epidemic	3	2	6
8.	Peculiarities of the time of the epidemic	3	1	3
9.	Unusually rapid spread of the epidemic	3	1	3
10.	Limitation of the epidemic to a specific population	3	1	3
11.	Peculiarities of the clinical manifestation	3	1	3
	Total			51

Table 6.10 Assessment of the Anthrax outbreak – USA 2001 with the Dembek et al. method (2007) (positive clues)

3.	Uncommon disease
5.	Multiple epidemics
9.	Unusual disease manifestation
11.	Direct evidence

Table 6.11 Assessment of the Anthrax outbreak – USA 2001 with the Radosavljevic & Belojevic method (2011)/indicators scored with 1/

No.	Epidemiological indicator
1.	Unusual/atypical manifestation of a known disease
3.	A sudden unexplainable increase in the number of cases or deaths in human populations
4.	Higher than expected morbidity and or mortality rates
5.	Clustering of patients with fever only or with fever and other symptoms
6.	A disease identified in the region for the first time again after a long period of time or after its eradication
7.	A new strain of pathogen identified in the region for the first time after a long period or after its eradication
8.	A disease with an unusual/atypical seasonal distribution
9.	One or more explosive epidemics/outbreaks with indicators of point-source origin
10.	A disease with an unusual geographic distribution.

Table 6.12 Assessment of the Anthrax outbreak – USA 2001 with the Radosavljevic method (2013)/indicators scored with 1/

No.	Perpetrator/Source of epidemic/Reservoir of pathogen	Biological agent/Pathogen	Means/Media/Factors of transmission	Target/Susceptible population
1.	Sophistication	A category (CDC)	Air	Intelligence
2.	Motivation	Amount of the pathogen	Fomites	Secrecy
3.	Intention		Delivery systems	Importance of a target/population
4.	Intelligence			
5.	Secrecy			
6.	Number of perpetrators			
7.	Number of sources of infection/reservoirs			
8.	Accessibility to sources of agent/pathogen			
9.	Accessibility to targets/ population at risk			

6.3.4 Anthrax- Sverdlovsk Soviet Union 1979

This outbreak of inhalational anthrax occured in April–May 1979. There were 66 reported human deaths and numerous deaths among animals in a narrow geographical zone of 4 km for humans and 40 km for animals. Later on it was revealed that accident happened at a military microbiological facility Compound 19, on 2 April 1979. There was a massive antibiotic distribution and vaccination among about 50.000 people [10].

Using the Grunow and Finke method in assessing this outbreak, a total score of 38 indicates that the intentional use of a biological warfare agent is likely (Table 6.13).

When the Dembek et al. method is applied on this outbreak, seven out of 11 clues are positive, indicating a high probability of an artificial outbreak (Table 6.14).

By the Radosavljevic & Belojevic method this UEE is assessed as a probably artificial outbreak as a total score is 9 (Table 6.15).

If the Radosavljevic method is used to assess this UEE, total score of 7 indicates that a biological attack is lowly probable and that an accident is more likely (Table 6.16).

Table 6.13 Assessment of the Anthrax outbreak – Sverdlovsk, Soviet Union 1979 with the Grunow & Finke method (2002)

No.	Criterion	Assessment (Points)	Weighting factor	Points
1.	Existence of a biological risk	2	2	4
2.	Existence of a biological threat	0	3	0
3.	Special aspects of a biological agent	2	3	6
4.	Peculiarities of the geographic distribution of the biological agent	3	1	3
5.	High concentration of the biological agent in the environment	3	2	6
6.	Peculiarities of the transmission mode of the biological agent	3	1	3
7.	Peculiarities of the intensity and dynamics of the epidemic	3	2	6
8.	Peculiarities of the time of the epidemic	3	1	3
9.	Unusually rapid spread of the epidemic	3	1	3
10.	Limitation of the epidemic to a specific population	1	1	1
11.	Peculiarities of the clinical manifestation	3	1	3
	Total			38

Table 6.14 Assessment of the Anthrax outbreak – Sverdlovsk, Soviet Union 1979 with the Dembek et al. method (2007)/positive clues/

No.	Clue
1.	A highly unusual event with large numbers of casualties
2.	Higher morbidity or mortality than expected
3.	Uncommon disease
4.	Point source outbreak
7.	Dead animals
9.	Unusual disease manifestation
10.	Downwind plume pattern

Table 6.15 Assessment of the Anthrax outbreak – Sverdlovsk, Soviet Union 1979 with the Radosavljevic & Belojevic method (2011)/indicators scored with 1/

No.	Epidemiological indicator
1.	Unusual/atypical manifestation of a known disease
3.	A sudden unexplainable increase in the number of cases or deaths in human populations
4.	Higher than expected morbidity and or mortality rates
5.	Clustering of patients with fever only or with fever and other symptoms
6.	A disease identified in the region for the first time again after a long period of time or after its eradication
7.	A new strain of pathogen identified in the region for the first time after a long period or after its eradication
8.	A disease with an unusual/atypical seasonal distribution
9.	One or more explosive epidemics/outbreaks with indicators of point-source origin
10.	A disease with an unusual geographic distribution.

Table 6.16 Assessment of the Anthrax outbreak – Sverdlovsk, Soviet Union 1979 with the Radosavljevic method (2013)/indicators scored with 1/

No.	Perpetrator/Source of epidemic/Reservoir of pathogen	Biological agent/Pathogen	Means/Media/Factors of transmission	Target/Susceptible population
1.	Intelligence	A category (CDC)	Air	Size of a target population
2.	Secrecy	Amount of the pathogen		
3.	Accessibility to sources of agent/pathogen			

6.3.5 West Nile Virus Outbreak – New York City, 1999

From August to October 1999 an outbreak of arboviral encephalitis ocurred in the New York City. There was an increase in bird deaths. West Nile Virus was isolated for the first time in Western Hemisphere. Intensive hospital-based surveillance confirmed 59 cases, including seven deaths in the area [2, 11]. By the Grunow and Finke method, it is assessed that the use of a biological warfare agent in this UEE is doubtful as a total score is 29 (Table 6.17).

If the Dembek et al. method is used to assess this UEE, four out of 11 clues are positive, indicating a low probability of a biological attack (Table 6.18).

When the Radosavljevic & Belojevic method is used to assess this UEE, a total score of 7 indicates a possibly artificial outbreak (Table 6.19).

Using the Radosavljevic method in assessing this UEE, a total score of 4 indicates a lowly probable biological attack. (Table 6.20).

Table 6.17 Assessment of the West Nile Virus outbreak, NYC 1999 with the Grunow & Finke method (2002)

No.	Criterion	Assessment (Points)	Weighting factor	Points
1.	Existence of a biological risk	3	2	6
2.	Existence of a biological threat	2	3	6
3.	Special aspects of a biological agent	0	3	0
4.	Peculiarities of the geographic distribution of the biological agent	3	1	3
5.	High concentration of the biological agent in the environment	2	2	4
6.	Peculiarities of the transmission mode of the biological agent	3	1	3
7.	Peculiarities of the intensity and dynamics of the epidemic	1	2	2
8.	Peculiarities of the time of the epidemic	1	1	1
9.	Unusually rapid spread of the epidemic	3	1	3
10.	Limitation of the epidemic to a specific population	0	1	0
11.	Peculiarities of the clinical manifestation	1	1	1
	Total			29

Table 6.18 Assessment of the West Nile Virus outbreak, NYC 1999 by Dembek et al. method (2007)/positive clues/

No.	Clue
1.	A highly unusual event with large numbers of casualties
2.	Higher morbidity or mortality than expected
3.	Uncommon disease
7.	Dead animals

Table 6.19 Assessment of the West Nile Virus outbreak, NYC 1999 by Radosavljevic & Belojevic method (2011)/indicators scored with 1/

No.	Epidemiological indicator
3.	A sudden unexplainable increase in the number of cases or deaths in human populations
4.	Higher than expected morbidity and or mortality rates
5.	Clustering of patients with fever only or with fever and other symptoms
6.	A disease identified in the region for the first time again after a long period of time or after its eradication
7.	A new strain of pathogen identified in the region for the first time after a long period or after its eradication
9.	One or more explosive epidemics/outbreaks with indicators of point-source origin
10.	A disease with an unusual geographic distribution.

Table 6.20 Assessment of the West Nile Virus outbreak, NYC 1999 by Radosavljevic method (2013)/indicators scored with 1/

No.	Perpetrator/Source of epidemic/Reservoir of pathogen	Biological agent/Pathogen	Means/Media/Factors of transmission	Target/Susceptible population
1.	Number of sources of infection/reservoirs	B category (CDC)	Vectors	Size of a target population

6.3.6 *Tularemia, Kosovo, 2000*

During October 1999 –May 2000 there was an outbreak of tularemia after a period of warfare. There were 900 suspected cases, and out of them 327 were confirmed. Many houses were left for months enabling a high number rodents to get in contact with food and water. The transmission was mainly foodborne. *Francisella tularensis* was serologically confirmed in diseased humans and in rodents [7, 20].

With the Grunow and Finke method it is assessed that the intentional use of a biological warfare agent in this UEE is doubtful as a total score is 19 (Table 6.21).

If the Dembek et al. method is used to assess this UEE four out of 11 clues are positive, indicating a low probability of a biological attack (Table 6.22).

When the Radosavljevic & Belojevic method is used to assess this UEE a total score of 2 indicates a naturally occured epidemic (Table 6.23).

Using the Radosavljevic method in assessing this UEE a total score of 8 indicates a lowly probable biological attack (Table 6.24).

Table 6.21 Assessment of the Tularemia outbreak, Kosovo 2000 by Grunow & Finke method (2002)

No.	Criterion	Assess. / Points/	Weight. factor	Tot
1.	Existence of a biological risk	1	2	2
2.	Existence of a biological threat	1	3	3
3.	Special aspects of a biological agent (BA)	0	3	0
4.	Peculiarities of the geographic distribution of BA	3	1	3
5.	High concentration of the BA in the environment	2	2	4
6.	Peculiarities of the transmission mode of BA	3	1	3
7.	Peculiarities of the intensity and dynamics of the epidemic	1	2	2
8.	Peculiarities of the time of the epidemic	0	1	0
9.	Unusually rapid spread of the epidemic	1	1	1
10.	Limitation of the epidemic to a specific population	0	1	0
11.	Peculiarities of the clinical manifestation	1	1	1
	Total			19

Table 6.22 Assessment of the Tularemia outbreak, Kosovo 2000 by Dembek et al. method (2007)/Positive clues/

No.	Clue
1.	A highly unusual event with large numbers of casualties
3.	Uncommon disease
5.	Multiple epidemics
9.	Unusual disease manifestation

Table 6.23 Assessment of the Tularemia outbreak, Kosovo 2000 by Radosavljevic & Belojevic method (2011)/indicators scored with 1/

No.	Epidemiological indicator
3.	A sudden unexplainable increase in the number of cases or deaths in human populations
6.	A disease identified in the region for the first time again after a long period of time or after its eradication

Table 6.24 Assessment of the Tularemia outbreak, Kosovo 2000by Radosavljevic method (2013)/ indicators scored with 1/

No.	Pepetrator/Source of epidemic/Reservoir of pathogen	Biological agent/Pathogen	Means/Media/ Factors of transmission	Target/Susceptible population
1.	Motivation	A category (CDC) (1)	Air	Size of a target population
2.	Number of sources of infection/reservoirs		Food	
3.			Water	
4.			Vectors	

6.3.7 Escherichia coli *O104:H4 Outbreak in Germany 2011*

From 1 May to 26 July 2011 Germany was hit by a large outbreak of diarrheal disease followed by hemolytic uremic syndrome caused by a new strain of enterohaemorrhagic *Escherichia coli* O104:H4. There were 2987 cases (18 lethal) of diarrhoea and 855 cases (35 lethal) of hemolitic uremic syndrome. Fenugreek sprouts grown from probably *Escherichia coli* contaminated seeds imported from Egypt were considered to be the most likely source of the outbreak. However, the possibility of an artificial outbreak should not be discarded [16, 17].

Using the Grunow and Finke method in assesing this outbreak a total score of 23 indicates that the use of a biological warfare agent is doubtful (Table 6.25).

When the Dembek et al. method is applied on this outbreak six out of 11 clues are positive, indicating a moderate probability of an artificial outbreak (Table 6.26).

With the Radosavljevic & Belojevic method this UEE is assessed as a probably artificial outbreak as a total score is 8 (Table 6.27).

If the Radosavljevic method is used to assess this UEE a total score of 11 indicates a possible biological attack (Table 6.28).

Table 6.25 Assessment of the Escherichia coli O104-H4 outbreak in Germany 2011 with the Grunow & Finke method (2002)

No.	Criterion	Assessment (Points)	Weighting factor	Points
1.	Existence of a biological risk	1	2	2
2.	Existence of a biological threat	0	3	0
3.	Special aspects of a biological agent	2	3	6
4.	Peculiarities of the geographic distribution of the biological agent	1	1	1
5.	High concentration of the biological agent in the environment	1	2	2
6.	Peculiarities of the transmission mode of the biological agent	1	1	1
7.	Peculiarities of the intensity and dynamics of the epidemic	2	2	4
8.	Peculiarities of the time of the epidemic	1	1	1
9.	Unusually rapid spread of the epidemic	2	1	2
10.	Limitation of the epidemic to a specific population	2	1	2
11.	Peculiarities of the clinical manifestation	2	1	2
	Total			23

Table 6.26 Assessment of the Escherichia coli O104-H4 outbreak in Germany 2011 with the Dembek et al. method (2007) /positive clues/

No.	Clue
1.	A highly unusual event with large numbers of casualties
2.	Higher morbidity or mortality than expected
3.	Uncommon disease
4.	Point source outbreak
5.	Multiple epidemics
9.	Unusual disease manifestation

Table 6.27 Assessment of the Escherichia coli O104-H4 outbreak in Germany 2011 with the Radosavljevic & Belojevic method (2011) /indicators scored with 1/

No.	Epidemiological indicator
1.	Unusual/atypical manifestation of a known disease
3.	A sudden unexplainable increase in the number of cases or deaths in human populations
4.	Higher than expected morbidity and or mortality rates
5.	Clustering of patients with fever only or with fever and other symptoms
7.	A new strain of pathogen identified in the region for the first time after a long period or after its eradication
8.	A disease with an unusual/atypical seasonal distribution
9.	One or more explosive epidemics/outbreaks with indicators of point-source origin
10.	A disease with an unusual geographic distribution.

Table 6.28 Assessment of the Escherichia coli O104-H4 outbreak in Germany 2011 with the Radosavljevic method (2013)/indicators scored with 1/

No.	Perpetrator/Source of epidemic/Reservoir of pathogen	Biological agent/Pathogen	Means/Media/Factors of transmission	Target/Susceptible population
1.	Sophistication	B category (CDC)	Food	Size of a target population)
2.	Motivation			
3.	Intention	Emerging pathogen		
4.	Number of reservoirs			
5.	Accessibility to agent sources	Amount of the pathogen		
6.	Accessibility to targets			

6.4 Discussion

We show that the four available methods for differentiation between a biological attack and other epidemics enable similar assessments when applied post hoc on different UEEs. There are obvious similarities between the methods and some indicators are common. The shorter questionnaires with 10–11 indicators and simple scoring are the characteristics of the Dembek et al. and the Radosavljevic and Belojevic methods. Therefore, these methods may be suggested during an epidemic even though it might be expected that the full application of the methods would not be possible due to insufficiency of data. However, they might be useful for a fast orientation. The Grunow and Finke method includes more complex scoring and may be recommended in the aftermath of an epidemic. The Radosavljevic method is the most detailed and allows for a differentiation between a biological attack, a spontaneous outbreak of a new or re-emerging disease, an accidental release of a pathogen and a natural outbreak of a known endemic disease that may mimic bioterrorism or biowarfare.

Specific epidemiogical tools for a fast orientation prior to laboratory analyses are helpful in UEEs because an insufficiency of available data and a lack of clinical experience may be expected related to a disease caused by a biological attack [3]. If some of these epidemiological tools warn of a possible biological attack, the precautionary principle may be applied and security forces engaged prior to a laboratory confirmation of a biological attack. It is important to act in time in order to neutralize possible perpetrators before they escape or prepare another attack. Together with a laboratory response network, these epidemiological tools in the hands of public health workers form the first health barrier against bioterrorism [19]. It is possible to prepare some of these epidemiological tools as a smartphone application in order to act as quickly as possible [1].

Early detection of a biological attack and a quick action against the perpetrators are the most efficient methods to reduce morbidity, mortality and economic damage. The analyzed four methods may be a powerful tool in these efforts. Beside urban areas that are usually in focus for preventive measures against bioterrorism, it is also important to train health care workers in rural areas using an interagency approach [5].

6.5 Conclusion

The four available methods are closely related in differentiation between biological attack and other epidemics. Although insufficiency of data might be expected during an epidemic, the simplier methods such as the Dembek et al. method and the Radosavljevic and Belojevic method may be useful for a fast orientation. The Grunow and Finke method is more complex and most effective after an epidemic. The Radosavljevic method is the most detailed and allows for a further differential analysis of an UEE. According to the precautionary principle if the results of these epidemiological methods pointed out to a highly probable bioterrorist attack, counter-terrorist measures should be applied prior to completing the microbial forensics which may be time consuming.

Acknowledgments This study was financially supported by the Serbian Ministry of Educatiuon, Science and Technological Development, Contract No. 175078.

References

1. Bachmann DJ, Jamison NK, Martin A, Delgado J, Kman NE (2015) Emergency preparedness and disaster response: there's an app for that. Prehosp Disaster Med 30:486–490
2. CDC (1999) Outbreak of West Nile like viral encephalitis – New York 1999. Morb Mortal Wkly Rep 48:845–849
3. Chang M, Glynn MK, Groseclose SL (2003) Endemic, notifiable bioterrorism-related diseases, United States, 1992–1999. Emerg Infect Dis 9:556–564
4. Dembek ZF, Kortepeter MG, Pavlin JA (2007) Discernment between deliberate and natural infectious disease outbreaks. Epidemiol Infect 135:353–371
5. Grundmann O (2014) The current state of bioterrorist attack surveillance and preparedness in the US. Risk Manag Health Policy 7:177–187
6. Grunow R, Finke EJ (2002) A procedure for differentiating between the intentional release of biological warfare agents and natural outbreaks of disease: its use in analyzing the tularemia outbreak in Kosovo in 1999 and 2000. Clin Microbiol Infect 8:510–521
7. Hsu VP, Lukacs SL, Handzel T et al (2002) Opening a bacillus anthracis-containing envelope, Capitol Hill, Washington, D.C.: the public health response. Emerg Infect Dis 8:1039–1043
8. Jernigan JA, Stephens DS, Ashford DA et al Bioterrorism Investigation Team (2001) Bioterrorism-related inhalational anthrax: the first 10 cases reported in the United States. Emerg Infect Dis 7:933–944
9. Kolavic SA, Kimura A, Simons SL, Slutsker L, Barth S, Haley CE (1997) An outbreak of Shigella dysenteriae type 2 among laboratory workers due to intentional food contamination. JAMA 278:396–398
10. Meselson M, Guillemin J, Hugh-Jones M et al (1994) The Sverdlovsk anthrax outbreak of 1979. Science 266:1202–1208

11. Mostashari F, Bunning ML, Kitsutani PT et al (2001) Epidemic West Nile encephalitis, New York, 1999: results of a household-based seroepidemiological survey. Lancet 358:261–264
12. National Research Council (US) (2002) Panel on biological issues. Countering bioterrorism: the role of science and technology. National Academies Press (US), Washington, DC
13. Radosavljevic V (2013) A new method of differentiation between a biological attack and other epidemics. In: Hunger I, Radosavljevic V, Belojevic G, Rotz L (eds) Biopreparedness and public health. Springer, Heidelberg, pp 17–32
14. Radosavljevic V, Belojevic G (2009) A new model of bioterrorism risk assessment. Biosecur Bioterror 7:443–451
15. Radosavljevic V, Belojevic G (2012) Unusual epidemic events: a new method of early orientation and differentiation between natural and deliberate epidemics. Public Health 126:77–81
16. Radosavljevic V, Finke EJ, Belojevic G (2015) Escherichia coli O104:H4 outbreak in Germany – clarification of the origin of the epidemic. Eur J Pub Health 25:125–129
17. Radosavljević V, Finke EJ, Belojević G (2016) Analysis of Escherichia coli O104:H4 outbreak in Germany in 2011 using differentiation method for unusual epidemiological events. Cent Eur J Public Health 24:9–15
18. Török TJ, Tauxe RV, Wise RP et al (1997) A large community outbreak of salmonellosis caused by intentional contamination of restaurant salad bars. JAMA 278:389–395
19. Wagar E (2016) Bioterrorism and the role of the clinical microbiology laboratory. Clin Microbiol Rev 29:175–189
20. World Health Organization (2000) Outbreak news. Wkly Epidemiol Rec 75:133

Chapter 7
Primordial and Primary Levels of Biothreat and Bioterrorism Prevention

Vladan Radosavljevic

Abstract There is still an abundance of preventive and solving measures against biological attacks that makes confusion and dezorientation among experts and health policy-makers. Our pyramidal model of adversaries, and spherical system of prevention help us to solve this problem. They make clearly to us, which measures should be applied at any of four levels of prevention without robust spending.

Primordial level of prevention should be focused to stop entering perpetrator/source of infection/reservoir of pathogen and biological agent/pathogen on defended territory. This is the first line of biodefense, deeply and multiply linked with the strategies of intelligence and deterrence.

The primary prevention of biological attack is focused on monitoring and surveillance of potential internal sources of biological agents and bioterrorists. We elaborate three types of surveillance: clinical (syndromic), laboratory and environmental.

Both levels of prevention were detailed analyzed, according to the next issues: ***Perpetrator/source of infection/reservoir of pathogen*** (Sophistication, Motivation, Intention, Intelligence, Secrecy, Number of perpetrators, Number of sources of infection/reservoirs, Accessibility to sources of agent/pathogen, Accessibility to targets/population at risk), ***Biological agent/pathogen*** (A category, B category, C category, Emerging pathogens, Amount of the available agent/pathogen), ***Means/media of delivery/factors of transmission*** (Air, Water, Food, Fomites, Vectors, Biological ammunition, Delivery systems, Dispersion systems **mechanism of release**), ***Target/susceptible population at risk*** (Intelligence, Secrecy, Personal control, Control of means/media of delivery/factors of transmission, Physical protection, Protection by chemoprophylaxis, Protection by immunoprophilaxis, Number of people in a target/population at risk, Importance of target/population at risk, Location of target/population, Distribution of people in a target/population at risk), and for each issue a whole spectrum of cheap, simple and effective preventive measures were proposed.

V. Radosavljevic (✉)
Military Medical Headquarter, Ministry of Defence, Crnotravska 17, Belgrade, Serbia

University of Defence, Belgrade, Serbia
e-mail: vladanr4@gmail.com

V. Radosavljevic et al. (eds.), *Defence Against Bioterrorism*, NATO Science for Peace and Security Series A: Chemistry and Biology,
https://doi.org/10.1007/978-94-024-1263-5_7

7.1 Introduction

To make clear our approaches and proposals within biodefense [1], we start with systematic, logical and concrete explanation of the key terms:

1. doctrine of biodefense,
2. strategy in biodefense and
3. investment in biodefense.

Doctrine in biodefense defines enemies, potential enemies and available resources in biodefense. Each country should define a doctrine for itself and then to find out its place in the collective doctrine on international level.

Strategy in biodefense defines types of biorisks (types of bioattacks), their structure and feasibility, levels and ways of their prevention [2].

Investment in biodefense, should be preferably in scientific community. The priority task of scientific community is to offer solutions – develop methods and models to reduce life losses, costs and risks, than just to demand more money, more vaccines, more drugs and so on.

Following practical reasons (saving human and material resources), we have analyzed different scenarios of unusual epidemiological events (UEEs) and biological attacks (BAs), designed and applied the new methods of outbreak analysis (early and quickly orientation, subtle and detailed differentiation) [3–7].

At first sight it is obvious that there are numerous possibilities for BAs [3, 8, 9], and consequently, it was almost impossible to assess risk of BAs. Because of that, we have divided BAs into four components, and components into 10 qualitative and 22 quantitative parameters [3]. If by analogy propose BA as an equivalent to outbreak, then these 32 parameters may consider as an equivalent to links in Vogralic's chain. So, elimination as many as possible BAs parameters, means reducing or elimination risk from BAs. This was a base for the new approaches:

1. Vulnerability analysis, allows targets classification, identification their *"locus minoris resistentiae"*, and constructive plans for their improvements.
2. Feasibility analysis, identify possible scenarios, possibilities for their prevention, costs and possible losses.
3. Bioterrorism risk assessment (BTRA) is a unique approach of the two former analyses allows biosecurity experts and policy-makers very clear strategy in biodefense – distinct directions with several steps [3].

Depending on necessities, mentioned approach may have dual purpose:

1 Retrograd analysis – to assess which biodefence scenario was possible and at which degree;
2 Anterograd analysis – to estimate outbreak (BA) risk for any target and make measures for its reducing or complete elimination.

From the aspect of biodefense the most important is *Anterograd analysis*,[1] that start from the component *Target/Population at risk*. If we have list of possible *Targets* then should classify them according to common threaten parameters from the method. Next step is reducing or complete elimination threatening parameters as many as possible. By this way we can approximatelly estimate outbreak risk for any target and make measures for its defence. These methods present a step forward in global biosecurity.

Target Classification
We recognize six groups of targets:

1. **Hard targets, strategic level**

The aim of a bioattack is to produce mass effects and incite mistrust in authorities, as well as panic and fear among the population. Even the use of biological weapons at a tactical level can cause losses with strategic dimensions. One case of Middle East Respiratory Syndrome (MERS), Severe Acute Respiratory Syndrome (SARS) or Ebola viral hemorrhagic fever is enough to cause catastrophic economic consequences. Since Western countries have intensive food production and centralized food industries, only one successful bioterrorist action may contaminate a huge amount of food and threaten the lives of thousands or hundreds of thousands of inhabitants.

2. **Soft targets, strategic level**

These targets encompass large territories and large populations [10, 11]. Classic bioattack on big cities simultaneously from the air (aerosols) or dozen(s) of terrorist infected with high contagious and lethal pathogen. Biological attacks cause two types of epidemics: epidemics of infectious disease and epidemics of fear and panic.

3. **Hard targets, operational level**

Consequences of the operational level attack could be of strategic importance. In the case of military/intelligence targets, security and stability of the state are endangered.

4. **Soft targets, operational level**

Attacks on "softer" targets (airports, railway stations, food production industries) have both: direct political and economic consequences.

---

[1] Regarding *Retrograd analysis* the most informative are components *Perpetrator/Source/Reservoir of infection* and *Target/Population at risk*. They have the highest number of parameters and they are of the most importance for security experts. For the scientific community the most attention should be focused on two components: *Agent/Pathogen* (new or re-emerging pathogen, artificial or possible weaponized pathogen), and *Means/media of transmission/ Factors of Transmission* because of medical and academic reasons (possible new modes of transmission, period of communicability and consecutive contrameasures). Component *Agent/Pathogen* and last three parameters (*Importance of target/population at risk, Number of people in a target/population at risk, Distribution of people in a target/population at risk*) from component *Target/Population at risk*, are most important for assessing possibility of deliberate outbreak (DO) on any concrete target.

5. **Hard targets, tactical level**

The most probable targets are highly prominent and protected institutions (government buildings, media centers) and people (politicians, scientists, high officers), but in sense of effects could be of strategic importance.

6. **Soft targets, tactical level**

"Soft targets" are ordinary people at public places (respiratory agents released in crowded and closed spaces like theaters, cinemas, sports events and political meetings).

7.2 Primordial Level of Prevention

There is still an abundance of preventive and solving measures against biological attacks that makes confusion and dezorientation among experts and health policy-makers. Our pyramidal model of adversaries and spherical system of prevention help us to solve this problem [10]. They make clearly to us, which measures should be applied at any of four levels of prevention without robust spending.

Primordial level of prevention should be focused on Perpetrator/source of infection/reservoir of pathogen and Biological agent/pathogen.

Elimination of *Sophistication* should be focused on top scientists who are at the same time both a source of agents and potential perpetrator. Highly sophisticated perpetrators are more probably to use highly dangerous agents (category A agents or emerging agents), attack without suicidal intent, keep themselves unknown and try to attack "hard" targets. Such perpetrator(s) carried out the U.S.A. anthrax attacks in 2001. Highly motivated but low sophisticated perpetrators (mainly poor terrorists or fanatics with suicidal intent) will probably use the more available category B agents and attack public ("soft") targets.

Scientific community should be monitored and bioresearch regularly assessed. It is highly recommendable better coordination between both scientific and intelligence community. Dual-use bioresearch demands strict control of employed staff and research material. The most secure way is successive chain of production with separate staff for every step. The main components of laboratory biosecurity are physical security (restrict access to unauthorized individuals), personnel security (individual screening), material control and accountability (transparency regarding working material), transport security (reliable information about packaging and carrier), information security (sensitive information protection from public release). In June 2015, India and the U.S.A. signed a new 10-year defense framework agreement, to work together in developing a lightweight protective suit effective in chemical and biological hazard environments [12]. With the advancement in the bioresearch field, regulations should be updated to minimize risks and independent committees of industry leaders, agency officials and academics should be appointed to design and reform regulations based on the risk assessments.

Motivation. Terrorists may behave on two ways. Some of them want to avoid attribution for a bioattack, others want to claim credit for it. People accidentally included in natural outbreaks (as a source/reservoir of infection) and look like perpetrators at first sight, are always afraid and cooperative during investigations. Also, natural source/reservoir of infection always completely behaves according to epidemiological characteristics (incubation, period of communicability). We recommend to check the passports of suspicious people on the borders, where they were (endemic areas, outbreak territory, suspicious contacts during travel) and carried out immediately health measures on them, if necessary (surveillance, isolation, chemo- or seroprophilaxis).

Intention. Control of incriminated and suspicious people on the borders like their travels in last 2 months (endemic areas, outbreak territory, incriminated (terrorist supporters) countries with biosafety labs levels 3 and 4, suspicious contacts during travel) and if necessary, carried out immediately health measures on them (surveillance, isolation, chemo- or seroprophilaxis).

Perpetrator/source of infection/reservoir of pathogen. Check the epidemiological situation in the country of origin or coming, especially for respiratory diseases. Detailed, complete, up-to-date follow-up of imported animals and plants as well as suspicious persons, from the place of origin until the border of importing/entry country.

Intelligence. Complete and up-to-date global network for: dangerous pathogens and diseases, labs facilities and experts.

Secrecy. Focus on two groups of people and two types of places. First, top scientist employed at the labs bio-safety level (BSL) 3 and 4, and their labs. Second, people from endemic areas and/or outbreak encompassed territories, suspected on infection from Category A agents or emerging pathogens.

Number of perpetrators. Border control of people from endemic areas or territories with patients/carriers, suspected on infection by A category agents or emerging pathogens.

Number of sources of infection/reservoirs. Eliminate them as soon as possible and as many as possible. People should warn to avoid endemic areas or other territories contaminated with agents from the A category and/or emerging pathogens. Provide health-security and legislative measures for people who plan to travel there. Improve monitoring of incriminated areas through World Health Organization (WHO), Physicians without borders, local authorities, epidemiology intelligence. Control passports on borders (visited countries). If person, according above mentioned is suspicious put him/her under health surveillance until maximum disease incubation plus 2 days.

Accessibility to sources of agent/pathogen. Priority should be on the labs BSL 3 and 4 and employed staff, as well as outbreak encompassed territories and endemic areas suspected on infection by Category A agents or emerging pathogens. Documents abound, about the desire by the Islamic terrorist groups, seeking to obtain biological agents in order to cause terror [13].

Accessibility to targets/population at risk. Identify *loci minoris resistentiae* in food chains, water supply systems and maximally secure them. Suspicious people and patients/carriers who enter on defended territory, keep isolated during incubation.

7.2.1 Biological Agent/Pathogen

A category/Emerging pathogens. Strengthen efforts to eradicate or eliminate diseases caused by *A category/Emerging pathogens*. Further, monitor experts who are able to make agents out of governmental control. Monitor dual-use research facilities.

B category/C category. Surveillance through WHO, Physicians without borders, local authorities and epidemiology intelligence for: patients, carriers, suspicious contacts and immediate environment. If necessary introduce: isolation, vaccino-, seroprophilaxis and other appropriate measures.

Amount of the available agent/pathogen. Additionally to strictly border control for both: people (written before) and animals for import (possible sources/reservoirs of *tularemia, anthrax, SARS, swine flu, avian flu*). For *haemorrhagic fevers* strictly border control when import rodents.

7.2.2 Means/Media of Delivery/Factors of Transmission

Air. The aerosol route may be used for strategic (large-scale) attack. Such pathogens are stable in aerosol and capable to be dispersed (5-17 μm particle size) [14]. So, scientific research should be focused on the means which can neutralize aerosols of micrometer size and security efforts on control of air traffic because of dispersion possibility of the agent (aerosol).

(I) Measures addressed to people:

1. Terrorist or suspicious. Surveillance through intelligence and epidemiology intelligence.
2. Tourists, immigrants and refugees. Surveillance through: WHO, Physicians without borders and local authorities. Epidemiology intelligence of: patients, carriers, suspicious contacts and immediate environment.
3. Scientist or other lab staff. Surveillance through intelligence and epidemiology intelligence.

(II) Measures addressed to potential *sources/reservoirs* of respiratory transmissible diseases (anthrax, haemorrhagic fevers, tularemia, plaque): immediate reporting and isolation of any suspicious case, suspicious contacts, carriers and immediate environment. Strictly border control when import rodents.

(III) Measures addressed to endemic areas or other incriminated territories. Surveillance through: WHO, Physicians without borders, local authorities, epidemiology intelligence of patients, suspicious contacts, carriers and immediate environment.

Water. Maximally secure water supply systems from intruders.

Food. Maximally secure food chains, especially storage and distribution centers, as well as production facilities for fruits, vegetables and dairy products (food without heating preparation).

Fomites. Early detection and immediate reporting for even suspicious case with Category A /Emerging pathogen. Incriminated fomites burned or disinfected.

Vectors. Low probably event. Recommendation for air traffic control is aerosol dispersion of repellents in airplanes even during their stay in endemic – incriminated areas.

Biological ammunition. It is low probably factor of agents/pathogens transmission.

Delivery systems. Those are very probably means of agents/pathogens transmission. Maximally secure water supply systems and food chains, apply automatic handling in postal offices (letters, packages) and enforce border control, especially for food.

Dispersion systems/mechanism of release. Airplanes control because of aerosol (agent) dispersion possibility (very probably event).

7.2.3 Target/Susceptible Population at Risk

Intelligence. Intelligence may be global and local. On local level are: personal control, electronic surveillance systems, local intelligence and observations possible targets (repeated visits by pedestrians or vehicles to the target). Further are: control of media such as air, food, water and fomites (office equipment, postal letters, packages). On global level the most important is networking of health and security services.

Secrecy. The impact of secrecy has been evident in the aftermath of the 2001 anthrax letters. The U.S. Postal Service and the CDC knew that the Brentwood postal facility in Washington, DC, was contaminated, they waited for 4 days before closing the facility and treating workers with antibiotics. During that time, one worker had died of anthrax, another was close to death, and two were gravely ill. Another example is China in 2003, when the government denied the SARS epidemic for 6 weeks, causing international alarm and spread of the disease.

Personal control. The highest standards of control must be applied to "hard" targets. Personal control includes physical control of people (their health status) and behavioral control (CV reviewing, control of suspected behavior, control of communications/contacts).

Control of means/media of delivery/factors of transmission. Control of water includes use of bottled water and permanent surveillance of central water supply systems. Food control should follow the principle "from the farm to the fork." In likely targets should incorporate biosensors. In "hard" targets air conditioning systems with gradually increasing air pressure is recommendable.

Parameters of protection are physical, chemical and immunological.

Physical protection. The simplest form of physical protection of people is advice them to remain indoors in response to a biological attack alarm. This will simply prevent transmission of biological agents. Much more sophisticated is the use of air conditioning systems or systems of increasing air pressure in different parts of buildings. Furthermore, UV radiation sources may be used as physical protection.

Protection by chemoprophylaxis. Mainly, chemical protection refers to use of antibiotics. It is a great logistical challenge, since it provides protection as long as the available stocks last.

Protection by immunoprophylaxis. Mass immunization programs require careful assess of potential risks and benefits. A nationwide smallpox vaccination, carried out in November 2002 in the U.S., was based on the idea of smallpox-infected Iraqis invading USA. It was 145 serious adverse cases among vaccinated persons (hospitalization, permanent disability, life-threatening illness and at least 3 deaths) [3].

Number of people in a target/population at risk. Since overcrowded targets are at highest risk, people must be advised to avoid them or spaciously safely distribute themselves. Potential targets should be well organized.

Importance of target/population at risk. Biological attacks cause epidemics of infectious disease and epidemics of fear and panic. The final and ultimate goal of bioattacks is political/ideological. Epidemic or pandemic of fear and panic spread much faster, could be much larger than epidemic of infectious disease and consequently more appropriate for reaching final goal. Physical disease in the target population is coming to be the second important objective.

Location of target/population and Distribution of people in a target/population at risk. Potential targets should be safely organized, well protected, located out of dense urban areas and with easily accessed roads. Targets and consequences could be direct or indirect. Killing people and destroying their health is direct target – consequence. Economic losses and political implications are indirect targets – consequences.

Bioattacks on the operational and tactical level may have consequences of strategic importance. In the case of government/military/intelligence targets, security and stability of the state are endangered. That is why political consequences could be prompt and enormous. "Soft targets" are ordinary people at public places (respiratory agents released in crowded and closed spaces like theaters, cinemas, sports events and political meetings). The importance of "bioshield" activities in food production facilities can not be overestimated. Since Western countries have intensive food production and centralized food industries, only one successful bioterrorist action may contaminate a huge amount of food and threaten the lives of hundreds of thousands of inhabitants. Such bioterrorist acts make people change their behavior for years, decades or even permanently. For example, when food is incriminated for a relatively short period, people may change their diet permanently. Because of fear from a bioattack, people can change or leave their jobs or residences, or avoid traveling to certain regions.

7.3 The Primary Level of Prevention

The primary prevention against biological attack should comprises monitoring and surveillance of potential internal sources of both: biological agents and bioterrorists. There are three types of surveillance: laboratory, clinical (syndromic) and environmental.

Routinely laboratory surveillance could be hold at biosafety level two (BSL2) labs. In cases of suspicion Europe, USA, Canada, Russia and some other countries are able to carry out diagnostic at BSL4 within 1 day.

The most important for clinical (syndromic) surveillance during detection bioterrorism event is to maintain a high level of suspicion among physicians (continual medical education and up-to-date information).

The most important for environmental surveillance is „*in focus detection* "(sampling from environmental source to detect and identify agent).

In the case of outbreak suspicion we propose our „*Remote detection*"(by questionaire).[2]

Perpetrator/source of infection/reservoir of pathogen. Several agents from Category A are present in many countries, mainly sparsely present or currently absent as indigenous diseases (tularemia, viral hemorrhagic fevers, botulism, anthrax, plague). Also, the outbreak could occur due to accidental infection during the biological weapons research. Special monitoring and preventive measures should be addressed to the next groups:

(1) Terrorist, disaffected groups and individuals; (2) Tourists, immigrants and refugees; (3) Laboratory staff from the BSL 3 and 4 labs.

Sophistication. Control measures should be focused on labs staff (BSL 3 and 4) and endemic areas with haemorrhagic fevers. Unsophisticated terrorist could infect himself deliberately in endemic areas with Ebola virus or viruses of other hemorrhagic fevers and goes to big cities during period of communicability to infect as

---

[2]**REQUEST FOR THE RESPONSIBLE AUTHORITIES (Remote outbreak detection by questionaire)**

1. Was there unusual/atypical manifestation (fulminant course) of a known disease? 2. Were there several unusual/unexplained syndromes coexisting in the same case without any other explanation? 3. Was there sudden unexplainable increase in the number of cases or deaths in human populations? 4. Was there higher than expected morbidity and/or mortality rates? 5. Was there clustering of patients with fever only or with fever and other symptoms? 6. Was disease identified in the region for the first time, again after a long period of time or after its eradication? 7. Was it new strain of pathogen identified in the region for the first time, after a long period or after its eradication? 8. Was disease with an unusual/atypical seasonal distribution? 9. Was there one or more explosive epidemics/outbreaks with indicators of a point-source origin?

10. Was disease with an unusual geographic distribution? 11. Was there existence of a biological risk? 12. Was there existence of a biological threat? 13. Was there high concentration of the biological agent in the environment? 14. What were peculiarities of the transmission mode of the biological agent? 15. Was there limitation of the epidemic to a specific population? 16. Was there lower attack rates in protected individuals? 17. Were there dead animals? 18. Was there reverse spread? 19. Is there direct evidence of deliberate outbreak?

many as possible people. They are with suicidal tendencies and it is one of the most horrible bioterrorism scenario.

Motivation/Intention. The most probably are Islamic terrorist, neonazists, disaffected sects or individuals. They have strong political or ideological motives and intentions for such acts.

Intelligence. Epidemiological intelligence presents ability to get true and timely information on global and local levels related to a biological attack (preferably about terrorist/suspicious persons and their activities/movements, type of pathogen, unusual disease occurrence and endemic/incriminated areas).

Secrecy. Secrecy comprises the capacity to keep activities clandestine before an attack and to keep perpetrators unknown after an attack. Period between deployment of a bioweapon and its effects could be long enough to give terrorist chance to escape. Strategic attack using viral respiratory bioagents (influenza virus, SARS virus, MERS virus) is highly possible and in such cases it would be difficult to distinguish between a natural disaster and a bioterrorist act. A clandestine biological attack with highly dangerous agents (anthrax, smallpox, viral hemorrhagic fevers) is possible, but will be detected easily and quickly because it is large-scale and with very unusual dangerous agents. As potential perpetrators military/intelligence forces prefer clandestine attack but terrorist groups/individuals prefer publicly confirmed attacks.

Number of perpetrators could be numerous particularly from disaffected groups. International network of security intelligence and epi-intelligence should identify and follow incriminated groups/persons.

Number of sources of infection/reservoirs could be numerous, especially for agents from categories B, C and sometimes emerging agents. Local health authorities are mandatory for identification, surveillance, reporting and elimination *sources/reservoirs of infection.*

Accessibility to sources of agent/pathogen exists, probably on many ways, especially for agents from categories B, C and emerging agents during outbreaks. Health authorities are mandatory for identification, surveillance and reporting of: patient(s), suspicious case(s) or carrier(s) with some agent(s) from any of the categories A, B, C or emerging agents.

Accessibility to targets/population at risk probably is easy and quickly because of presence of both on the territory: agent/pathogen and possible perpetrator. Health service is obligatory for diagnostics and reporting about patient, suspicious case or carrier infected by agent(s) from any of the categories A, B, C or emerging agents/pathogens.

7.3.1 Biological Agent/Pathogen

Measures should be addressed to endemic areas or other incriminated territories to eradicate pathogen from the territory. In case that is not possible, surveillance depending on type of disease, reservoirs/sources of infection and means/media of transmission. Special attention should be addressed to respiratory diseases (hemorrhagic fevers).

A category/Emerging pathogens. There are two sources of infection/reservoirs of pathogen. The first one, are top scientists and other staff from the labs with BSL 3 and 4 from the own territory, and another one are endemic areas particularly with respiratory hemorrhagic fevers (own territory). Focus should be on control staff from the labs with BSL 3 and 4 and on travelers from endemic areas preferably with respiratory hemorrhagic fevers. Surveillance by WHO and other international health organizations, governmental health authorities is an asset.

B category and C category. State's health authorities are mandatory for surveillance, identification and reporting even suspicious case of disease from these categories of agents. Early detection could save many lives by triggering an effective containment strategy (isolation, chemo- and immunoprophilaxis). A developed network of data collecting, rapid data transmission to the relevant public health institutions and their careful analyses are priorities. Ultimate aim is to notice subtle differences between usual and unusual occurrence of diseases.

Amount of the available agent/pathogen. From small to mediate amounts of pathogen is possible to get disaffected persons, criminals or terrorists. But consequences could be of strategic importance.

7.3.2 Means/Media of Delivery/Factors of Transmission

Any kind of medium could be used. Means of delivery depend on the characteristics of the agent.

Air. The only mean of delivery for respiratory agents for strategic and clandestine use is airplane. The Congressional Office of Technology Assessment estimated that the aerosolized release by airplane of 100 kg of anthrax spores upwind of Washington, DC, could result in approximately 130,000 to 3 million deaths [15]. Dissemination of an agent through a ventilation/air conditioning system is another powerful way of attack by air. So, air traffic control is of the highest priority.

Food/Water. Centralized food production and water supply systems in developed countries increase vulnerability to foodborne [16, 17] and waterborne pathogens (most dangerous diseases are botulism and anthrax) [18, 19]. The most likely, terrorist groups and disaffected individuals could use drinking water and food for contamination with bioagents/pathogens. Multiple means of delivery are also possible; for example, anthrax can be an airborne or foodborne agent. Permanent monitoring and frequent control of centralized food production and water supply systems are mandatory.

Fomites. Could be used on tactical level but with strategic consequences (Ameritrax attack in 2001). May be carried out by states' institutions, such as military forces, intelligence services, well-funded organizations or individuals. Handling with fomties of mass using should be automatic.

Vectors. Means of delivery might be suicidal biobombers infiltrated in the targets, animals (birds infected with avian influenza, pigs infected with swine influenza and insects). Veterinary control of animal import has to be carried out completely.

Biological ammunition. May be produced and used by government or its institutions. Intelligence activities on suspicious persons are mandatory, as well as surveillance from public health institutions on reservoirs/sources of incriminated pathogens.

Delivery systems. Priority should be given security of water and food supply systems/chains. Postal delivery should be under monitoring and automatic as much as possible.

Dispersion systems/mechanism of release. Emphasize surveillance of air traffic control and public health surveillance especially above urban areas. Intelligence controls of suspicious persons are mandatory, as well as surveillance from public health institutions of reservoirs/sources of incriminated pathogens.

7.3.3 Target/Susceptible Population at Risk

Intelligence. Depending on both: existing sources/reservoirs of infection and possible perpetrators, should be organized and applied eradication programs, elimination programs and surveillance services for them. The intentional spread of anthrax in the USA has led to a surge in the development system able to integrate data from multiple sources into a single surveillance system oriented towards detection of unusual diseases, spread in unusual ways (continual systematic collection, analysis, interpretation and dissemination of data) [20]. Syndromic surveillance is monitoring clinical manifestations of certain illnesses. Laboratory surveillance comprises looking for certain laboratory data or biological markers. Environmental surveillance is the process by which the environmental samples are systematically analyze for the presence of biological agents [21].

Bioterrorism surveillance systems require three key features: timeliness, high sensitivity and specificity, and routine data analysis. Traditional biosurveillance system is based on the recognition and alert of a clear increase in diagnosed/suspected cases (to notice subtle differences between usual and unusual occurrence of diseases). For early detection of deliberate outbreaks the sensitivity of the systems need to be as high as possible [22]. Such system is inexpensive, simply to implement, free of technologic barriers and important component of global biosurveillance. Should be used together with methods which quickly identify the treat and institute public health protection measures (immunization, chemoprophylaxis and isolation).

Secrecy. Secrecy is an imperative to the authorities during bioattack and must be very well balanced (not to endanger public health but to mitigate fear and panic).

Simultaneously both, outbreak of infectious disease and epidemic of fear and panic could be caused by biological attack [10]. The main aims of bioterrorists are propagation of: fear, anxiety, uncertainty, depression of the population, mistrust in government and economic damage. An epidemic (or pandemic) of fear and panic multiplies the economic damage (losses in tourism, traffic, investment and export). Causation of physical disease is the second important objective. The final and ultimate goals of bioterrorists are political concessions. Reforming state public health legislation should be addressed to support such potential states' secrecy.

Personal control. For "hard" targets the recommend the same procedure as in Primordial prevention. For "soft" targets should apply "mass gathering medicine" preventive measures (improve security aspect of: selection of location, season (time) of event and control participants as many as possible especially from abroad, suspected groups and individuals).

Control of means/media of delivery/factors of transmission. Experts in bioterrorism should be involved in both, developing and maintaining surveillance systems. Innovative analytical methods should be able to provide interpretations of the data for: the spread of the outbreak, the identification of the source and early detection of outbreak [7].

Physical protection. Physical isolation of existing sources/reservoirs of infection, infected/diseased persons, carriers, suspected on infection individuals, and epidemiological/security surveillance of possible perpetrators.

Protection by chemoprophylaxis and by immunoprophylaxis. Should be reconsidered strategy that must have enough amounts of drugs and vaccines to every inhabitant. If the public health system is well developed, it is enough to have chemoprophylaxis for several dozens thousands inhabitants and immunoprophylaxis for several thousand inhabitants, for the first respond.

Importance of target/population at risk. Focus should be on finding out and monitoring both: existing internal sources/reservoirs of infection and possible internal perpetrators. If we are well introduced and control those issues then, there are both, enough time and enough information to be prepared for primary level of prevention, especially for "hard" targets.

Location of target/population at risk. Potential "soft" and "hard" targets, should be located to easy accessible highways and hospitals.

Number of people in a target/population at risk. Risky times request avoiding risky behaviors. Should avoid mass gathering. Strict control of food and water supply and delivery systems. In case of diseased people forbid mass gathering and isolate suspicious cases/contacts.

Distribution of people in a target/population at risk. The risk is very high in camps or every kind of temporary/overcrowded accommodation. In case of suspicious or confirmed patient/contact immediately carry out isolation. Recommendation is, as less as possible people in every room.

References

1. Radosavljevic V (2016) Biodefense system – communicable diseases and public health. J Bioterror Biodef 7:e121. https://doi.org/10.4172/2157-2526.1000e121
2. Radosavljevic V (2012) Strategies in fighting bioterrorism. J Bioterr Biodef 3:e1012. https://doi.org/10.4172/2157-2526.1000e102
3. Radosavljevic V, Belojevic GA (2009) new model of bioterrorism risk assessment. Biosecur Bioterror **7**:443–451
4. Radosavljevic V, Belojevic G (2012) Unusual epidemilogical event – new model for early orientation and differentiation between natural and deliberate outbreak. Public Health 126(1):77–81
5. Radosavljevic V, Finke E-J, Belojevic G (2015) Escherichia coli O104:H4 outbreak in Germany—clarification of the origin of the epidemic. Eur J Public Health 25(1):125–129
6. Radosavljevic VA (2013) new method of differentiation between a biological attack and other epidemics. In: Hunger I, Radosavljevic V, Belojevic G, Rotz L (eds) Biopreparedness and public health. Springer, Heidelberg, pp 17–32
7. Radosavljevic V, Finke E-J, Belojevic G (2016) Analysis of the Escherichia coli O104:H4 outbreak in Germany in 2011 by a differentiation method for unusual epidemiological events. Cent Eur J Public Health 24(1):9–15
8. Radosavljevic V, Stojkovic K, Anđelkovic R, Andrejic M (2010) Agroterrorism as actual challenge. Vojnosanit pregl 67(11):933–940
9. Radosavljevic V, Belojevic G, Jovanovic L (2012) A mathematical model of bioterrorist attack risk assessment. J Bioterr Biodef 2:114. https://doi.org/10.4172/2157-2526.1000114
10. Radosavljevic V, Jakovljevic B (2007) Review: Bioterrorism – types of epidemics, new epidemiological paradigm and levels of prevention. Public Health 121:549–557
11. Radosavljevic V, Radunovic D, Belojevic G (2009) Epidemics of panic during a bioterrorist attack – a mathematical model. Med Hypothesis 73:342–346
12. Garamone J (2015) U.S., India sign 10-year defense framework agreement," US Department of Defense. www.defense.gov
13. Galamas F (2011) Profile bioterrorism: present and potential threats. Comp Strateg 30:70–93
14. Sinha S, Singh J (2016) Classification, causes, control measures and acts of bioterrorism. Int J Appl Biol Pharm Technol 7(2):342–354
15. Inglesby TV, Henderson DA, Bartlett JG et al (1999) Anthrax as a biological weapon: medical and public health management. Working group on civilian biodefense. JAMA 281:1735–1745
16. Alberts B (2005) Modeling attacks on the food supply. Proc Natl Acad Sci U S A 102:9737–9738
17. Wein LM, Liu Y (2005) Analyzing a bioterror attack on the food supply: the case of botulinum toxin in milk. Proc Natl Acad Sci U S A 102:9984–9989
18. Nuzzo J (2006) The biological threat to U.S. water supplies: toward a national water security policy. Biosecur Bioterror 4:147–159
19. Burrows WD, Renner SE (1999) Biological warfare agents as threats to potable water. Environ Health Perspect 107:975–984
20. Nicholas E, Kman Daniel J (2012) Bachmann biosurveillance: a review and update. Adv Pre Med 2012. doi:https://doi.org/10.1155/2012/301408
21. Radosavljevic V (2011) Environmental health and bioterrorism. In: Nriagu JO (ed) Encyclopedia of environmental health, vol 2. Elsevier, Burlington, pp 392–399
22. Steinbrook R (2005) Biomedical research and biosecurity. New Engl J Med 353:2212–2214

Chapter 8
Secondary Level of Biothreat and Bioterrorism Prevention

Vladan Radosavljevic

Abstract Secondary level of bioterrorism/outbreak prevention is third level of prevention and carries out when diseased confirmed. It is the most emerging and the most demanding part of biothreat&bioterrorism prevention. As a help to common citizen, response should be addressed to classic antiepidemic measures and to mitigation fear and panic.

8.1 Introduction

This is the most emerging and the most demanding part of outbreak/bioterrorism act. Measures of secondary prevention should be addressed to breaking both: epidemic of infectious disease and the epidemic of fear and panic [1]. Consequently, response has to comprise public health scenario and security scenario. The public health scenario manages the disease caused directly by the bioattack/outbreak. The security scenario involves dealing with fear, panic, health concerns and other psychological reactions that normally arise in disasters [2].

8.1.1 Perpetrator/Source of Infection/Reservoir of Pathogen

If this component is determined must be immediately eliminated, if it is still unknown its discovering must be priority [3–7]. Network of three pillars (preventive medicine, intelligence and security) should be trained and ready, on both: state and international level. Investigation should begin from the local and central authorities (teams). Investigator team(s) should send focused questions as soon as possible to local authorities: "Was there one or more explosive epidemics/outbreaks with

V. Radosavljevic (✉)
Military Medical Headquarter, Ministry of Defence, Crnotravska 17, Belgrade, Serbia

University of Defence, Belgrade, Serbia
e-mail: vladanr4@gmail.com

V. Radosavljevic et al. (eds.), *Defence Against Bioterrorism*, NATO Science for Peace and Security Series A: Chemistry and Biology,
https://doi.org/10.1007/978-94-024-1263-5_8

indicators of a point-source origin? Is there direct evidence?" (See Chapter: "The primary level of prevention").

Sophistication Accurate assessment of sophistication could significantly help in differentiating type of outbreak [8]. (What were peculiarities of the transmission mode of the biological agent? Was there unusual/atypical manifestation (fulminant course) of a known disease? Were there several unusual/unexplained syndromes coexisting in the same case without any other explanation?)

Motivation/Intention These issues are informative, but in some cases could be used malicious as mimicry (Was there existence of a biological risk? Was there existence of a biological threat?).

Intelligence Depending on type of outbreak, check all contacts with *source of infection/reservoir of pathogen,* even though possible indirect contacts [9]. Especially be careful if it is disease with an unusual/atypical seasonal distribution and/or with an unusual geographic distribution.

Secrecy If bioterrorist attack is not secret, then will be discovered and prevented. Early detection and diagnosis of suspicious cases as well as trained network of three pillars (preventive medicine, intelligence and security) should be carried out, on both: state and international level [10, 11].

Number of Perpetrators Even one perpetrator may produce bioterror-attack with strategic significance and consequences. Intelligence and security forces should be on alert, on a state and international level.

Number of Sources of Infection/Reservoirs/Accessibility to Sources of Agent/Pathogen Priority of each government should be elimination/eradication sources/reservoirs of infection, because of health and security reasons. It is a first task for preventive medicine. Because, only one case of severe infective disease, very easily could become an international health and security problem, this emphasizes necessity for global networking and teams for emergency response.

Accessibility to Targets/Population at Risk It is necessary to break any of the four components of bioattack/outbreak. This is a new, elegant and very approachable methodology for that task.

8.1.2 Biological Agent/Pathogen

A Category There are only six different pathogens in this category, and relatively easy could be determined nature of the agent (natural or weaponized). This highly informative indicator may help significantly in clarification of the outbreak origin

[12]. Before arriving expert team(s) in incriminated area, next question should be asked:" Was it new strain of pathogen identified in the region for the first time, after a long period or after its eradication?" It could be a large logistic challenge because of necessity of massive vaccine- and chemoprophylaxis. Promptness and complete antiepidemic measures are crucial.

B Category and C Category Terrorist could be easily infected and spread pathogens among susceptible population. Also, it is highly likely to be natural or reemerging outbreak. Anyway, prompt, efficient contingency plans with high logistic capacities should exist (large amounts of vaccines and chemoprophylaxis drugs).

Emerging Pathogen This group of pathogens may cause natural, deliberate, new/reemerging and accidental type of outbreak. Before arriving expert team(s) in incriminated area, should be asked:" Was disease identified in the region for the first time, again after a long period of time or after its eradication?"

Amount of the Available Agent/Pathogen The next question should be asked as soon as possible: "Was there high concentration of the biological agent in the environment?" If the answer is "Yes", then there is a high probability of deliberate or accidental caused outbreak. In such case security services must be immediately alerted.

8.1.3 Means/Media of Delivery/Factors of Transmission

Air It is the most dangerous way of disease transmission and very difficult to control (isolation, vaccino- and chemoprophylaxis, epidemiological surveillance).

Food and Water In developed countries with centralized food and water production as well as centralized food and water supply, food and water are very probable ways of disease transmission. This is task for sanitary and security services [13–16].

Fomites The most dangerous is possibility of their use to attack "hard" targets (Ameritrax in 2001).

Vectors The next questions should be asked as soon as possible: "Were there dead animals? Was there reverse spread?" This way of disease transmission is usually used for "soft" targets. Could be agroterroristic act directed to both, animals and people if antropozoonotic disease in case [17, 18]. Veterinary services must be included in solving outbreak.

Biological Ammunition It is conclusive indicator of biological attack, but must be considered with cautious because of possibility to be false imputed.

Delivery Systems Could be misused as way(s) of transmission (Ameritrax in 2001, German *E. coli* outbreak in 2011) [19, 20]. So, strict control from all aspects (personal checking, hygienic conditions) in food delivery systems and water supply systems are necessary.

Dispersion Systems/Mechanism of Release This is the most dangerous way of transmission for air-borne, food-borne and water-borne diseases. Could be accidental (German *E. coli* outbreak in 2011) or deliberate (Ameritrax in 2001). Air traffic control, control of international passengers especially from incriminated areas with respiratory diseases and control of food and water delivery and storage centers.

8.1.4 Target/Susceptible Population at Risk

Intelligence The next information should be provided as soon as possible: "Was there sudden unexplainable increase in the number of cases or deaths in human populations? Was there higher than expected morbidity and/or mortality rates? Was there clustering of patients with fever only or with fever and other symptoms? Was there limitation of the epidemic to a specific population? Was there lower attack rate in protected individuals?"

Secrecy If data about outbreak are incomplete or suspicious, must not be spread. Manage of information should avoid rumors, panic, unnecessary fear and confusion.

Personal Control Emphasized control should be addressed on two groups of people: (1) suspicious or confirmed terrorist and (2) people employed in food and water supply, air traffic control and "hard" targets. Their contacts, behavior, dedication to service must be checked regularly.

Control of Means/Media of Delivery/Factors of Transmission Strictly control should be addressed on both food and water production and supply systems.

Physical Protection Should provide hospital facilities in reserve (enough hospital rooms and toilets), like abandoned barracks, resting places, schools, etc.

Protection by Chemoprophylaxis Provide large quantities of antibiotics, antipyretics, analgetics, antidiarrheal drugs, infusion solutions for rehydratation, etc.

Protection by Immunoprophylaxis During outbreak seroprophylaxis is mandatory for people who were in contacts with: diseased, suspected on serious infection, suspected contacts and vulnerable groups (elderly, chronically ill, pregnant, children). Vaccination is mandatory usually for the whole country in case of serious infective disease. For majority vaccines and majority of populations 2 weeks are enough to develop full vaccine immunity after vaccination. Even in few days vaccine may develop immunity, not in full capacity, but very useful to save a life in combination with symptomatic or causal therapy. Vaccination has two major limitations. First, allergic reactions on its components in some individuals. In such cases, it is matter of epidemiologist/medical doctors to assess possibility of vaccination (to do desenzibilization allergic person) or do some other protective measures. Second, even vaccinated many persons may be infected and get disease at different level of seriousness (mild, middle, heavy clinical picture). If infectious dose is big, people in good health conditions may diseased and died (brakethrough of vaccine immunity). So, besides of immunoprophylaxis, all other prevention measures must be carried out in case of serious outbreak.

Importance of Target/Population at Risk During outbreak two epidemics are occurs simultaneously: outbreak of infective disease and outbreak of fear and panic. There are a lot of malevolent sources which are keen to "pump" fear and panic in threaten population. High intensity of panic and fear are obstacles, and in the worst cases may block antiepidemic and other public health measures. For complete antiepidemic measures are highly important appropriate manage of both: public opinion and public information.

Even one case of very contagious and deadly disease is highly important for target/population at risk on country/national level. If case that dangerous infective disease confirmed, public should be informed, health system on alert and surveillance must last until the case is solved and incubation period passed (plus 2 days) in persons who were in contact with diseased.

Location of Target/Population at Risk Regarding modern traffic dissemination airborne diseases are particular, permanent threat. "Soft" targets, like urban areas and big cities are at highest risk, and from their preparedness and readiness to prevent and defeat outbreak mainly depends destiny of the rest of the country.

Number and Distribution of People in a Target/Population at Risk Big, urban, densely populated areas are most "attractive" and "grateful" targets for the terrorists. Consequently, they should be pivotal in biodefence. Successful or partially successful bioattack in such areas is of global importance in the field of security-antiterrorism and public health, and it is recommendable to be solved by common international efforts [21, 22].

References

1. Radosavljevic V, Jakovljevic B (2007) Review: bioterrorism – types of epidemics, new epidemiological paradigm and levels of prevention. Public Health 121:549–557
2. Radosavljevic V, Radunovic D, Belojevic G (2009) Epidemics of panic during a bioterrorist attack – a mathematical model. Med Hypothesis 73:342–346

3. Radosavljevic V, Belojevic G (2009) A new model of bioterrorism risk assessment. Biosecur Bioterror 7:443–451
4. Radosavljevic V, Belojevic G, Jovanovic L (2012) A mathematical model of bioterrorist attack risk assessment. J Bioterr Biodef 2:114. https://doi.org/10.4172/2157-2526.1000114
5. Radosavljević V, Belojević G (2011) Biološko oružje – novi pristupi u borbi protiv bioterorizma (Biological weapons – the new approaches in fighting against bioterrorism). Medija centar „Odbrana“, Beograd
6. Radosavljevic V, Belojevic G (2012) Unusual epidemilogical event – new model for early orientation and differentation between natural and deliberate outbreak. Public Health 126:77–81
7. Radosavljevic V (2013) A new method of differentiation between a biological attack and other epidemics. In: Hunger I, Radosavljevic V, Belojevic G, Rotz L (eds) Biopreparedness and public health. Springer, Heidelberg, pp 7–32
8. Galamas F (2011) Profile bioterrorism: present and potential threats. Comp Strateg 30:70–93
9. Sinha S, Singh J (2016) Classification, causes, control measures and acts of bioterrorism. Int J Appl Biol Pharm Technol 7:342–354
10. Garamone J (2015) U.S., India sign 10-year defense framework agreement," US Department of Defense. www.defense.gov
11. Steinbrook R (2005) Biomedical research and biosecurity. N Engl J Med 353:2212–2214
12. Inglesby T, Henderson D, Bartlett J et al (1999) Anthrax as a biological weapon: medical and public health management. Working group on civilian biodefense. JAMA 281:1735–1745
13. Alberts B (2005) Modeling attacks on the food supply. Proc Natl Acad Sci U S A 102:9737–9738
14. Wein L, Liu Y (2005) Analyzing a bioterror attack on the food supply: the case of botulinum toxin in milk. Proc Natl Acad Sci U S A 102:9984–9989
15. Nuzzo J (2006) The biological threat to U.S. water supplies: toward a national water security policy. Biosecur Bioterror 4:147–159
16. Burrows W, Renner S (1999) Biological warfare agents as threats to potable water. Environ Health Perspect 107:975–984
17. Radosavljevic V (2011) Environmental health and bioterrorism. In: Nriagu JO (ed) Encyclopedia of environmental health, vol 2. Elsevier, Burlington, pp 392–399
18. Radosavljevic V, Stojkovic K, Anđelkovic R, Andrejic M (2010) Agroterrorism as actual challenge. Vojnosanit Pregl 67:933–940
19. Radosavljevic V, Finke E-J, Belojevic G (2015) Escherichia coli O104:H4 outbreak in Germany—clarification of the origin of the epidemic. Eur J Pub Health 25:125–129
20. Radosavljevic V, Finke E-J, Belojevic G (2016) Analysis of the escherichia coli O104:H4 outbreak in Germany in 2011 by a differentiation method for unusual epidemiological events. Cent Eur J Public Health 24:9–15
21. Radosavljevic VR (2016) Biodefense system – communicable diseases and public health. J Bioterror Biodef 7:e121. https://doi.org/10.4172/2157-2526.1000e121
22. Radosavljevic VR (2012) Strategies in fighting bioterrorism. J Bioterr Biodef 3:e1012. https://doi.org/10.4172/2157-2526.1000e102

Chapter 9
Preventative Medicine: Research and Use of Medical Countermeasures During an Outbreak

Inger Damon

Abstract New viral and bacterial pathogens are continuously emerging; today's world of increasing interconnectivity and mobility accelerates this shared global risk. As such, emerging infectious diseases will continue to require attentive public health surveillance programs in order to identify these events and allow effective responses and interventions to prevent or slow the spread of these diseases. Research that permits a better understanding of the potential source of infection, and the benefit of public health interventions such as therapeutics, vaccines, and diagnostics are critical components of preparedness. As an example, preparedness of smallpox medical countermeasures enabled a quick response to related monkeypox disease upon its introduction in the U.S. in 2003. In turn, the monkeypox response efforts identified a number of research questions that informed our understanding of monkeypox pathogenesis, and provided an animal model to evaluate smallpox antivirals and next generation vaccines. During the response to the Ebola Zaire virus epidemic in West Africa during 2014–2016, approaches that had been used to contain previous smaller outbreaks in DRC and Uganda proved successful in controlling disease spread and eventually, halted the epidemic. The size of the Ebola epidemic in West Africa provided unique opportunities to understand further how this virus affects the human host, and the persistence of virus in certain areas of the body. Additionally, Ebola medical countermeasures, evaluated later in the response, may have application in ongoing prevention efforts and in diminishing the size of future outbreaks. Evaluation or observational studies carried out during a response are important in that they provide evidence of best practices. This essay, a synopsis of a presentation made in Belgrade in March 2017, will expand on these specific examples of how

The findings and conclusions in this report are those of the authors and do not necessarily represent the views of the Centers for Disease Control and Prevention/the Agency for Toxic Substances and Disease Registry.

I. Damon (✉)
Division of High-Consequence Pathogens and Pathology National Center for Emerging and Zoonotic Infectious Diseases, Centers for Disease Control and Prevention, Atlanta, GA, USA
e-mail: iad7@cdc.gov

V. Radosavljevic et al. (eds.), *Defence Against Bioterrorism*, NATO Science for Peace and Security Series A: Chemistry and Biology,
https://doi.org/10.1007/978-94-024-1263-5_9

research can affect the outbreak response, and in turn, how effective evaluation during a response identifies new research questions.

9.1 Smallpox

Smallpox is a disease caused by the orthopoxvirus, variola virus. Smallpox major had an average case fatality rate of ~30%, whereas minor forms of the disease had a <1% case fatality rate. The disease was associated with a long incubation period 7–17 days, average 10–12 days, followed by the onset of fever and flu-like symptoms of headache, back ache, abdominal pain and or myalgias. Shortly thereafter, rash developed, and an individual would become infectious. Rash developed in characteristic stages, from macule, papule, vesicle, and papule to crust. Transmission to others was associated with close contact and likely through respiratory droplets. Secondary attack rates were highest in households, healthcare workers, and to those not previously vaccinated. No zoonotic sources of disease are known.

Surveillance, identification and isolation of cases, in addition to vaccination efforts in the mid-twentieth century led to the WHO declaration of smallpox eradication in 1980 [3]. Routine smallpox vaccination, using the related orthopoxvirus known as vaccinia virus, ceased after eradication of smallpox. Smallpox vaccines used viruses characterized to create localized lesions (the "take") and the origin of vaccinia virus remains enigmatic [2].

Although smallpox disease was declared eradicated, focus on a smallpox research agenda was augmented in 1999 after the Institute of Medicine (IOM) released its report, "Future Needs for Live Variola Virus", to guide advanced research and development of smallpox medical countermeasures. Increased concerns about the potential for bioterror use of variola virus led to the development of a research agenda associated with preparedness needs including additional genomic and biologic characterization of variola viruses in order to develop sensitive and specific diagnostics, and development of effective antivirals and vaccines with less reactogenicity for those with compromised immune systems. Additionally, the report outlined that in the absence of human disease, it would be important to develop well-characterized biologic surrogate systems that could be used to demonstrate efficacy of antivirals.

Between 1999 and 2003, research focused on the sequencing of variola viruses as well as screening of antiviral compounds and the initial characterization of a non-human primate model of smallpox [15]. The accomplishments of the smallpox research agenda ultimately led to the development of diagnostics, therapeutics and less reactogenic vaccines and are summarized in a few reviews [4]. Clinical approaches to the use of smallpox vaccines in an emergency have also been summarized [12].

Sequencing efforts that included near neighbor viruses led to the development of a non-variola orthopoxvirus diagnostic to be used to identify an orthopoxvirus that was not variola. This assay was distributed to the Laboratory Response Network (LRN), a consortium of public health laboratories in the U.S., in order to support healthcare worker smallpox vaccination efforts in 2001–2002. Subsequently, these assays were rapidly evaluated for monkeypox and enabled screening of rash specimens in the U.S. in 2003.

9.2 Monkeypox

Monkeypox is a non-variola orthopoxvirus that was identified as a human illness during the intensified surveillance efforts to certify the eradication of smallpox in West and Central Africa in 1970. It had initially been identified as a cause of rash illness in non-human primates. Monkeypox is a zoonotic disease; the reservoir is believed to be a rodent species. To date, the only isolate of monkeypox from a wild-caught animal was from a rope squirrel in 1986 [10].

Human monkeypox illness is very similar to that of smallpox, with the exception that lymphadenopathy is a prominent feature of monkeypox, and the case fatality rate is ~10% in children under the age of 5, and in those who had not been vaccinated between 3 and 19 years prior. Disease surveillance and epidemiology studies suggested that interhuman transmission was far rarer than that seen with smallpox, although some human-to-human transmission occurred within close household contacts. Most infections were in those who had primary exposure through hunting or preparation of meat from a variety of wildlife species.

In the early summer of 2003, the Marshfield clinic in Wisconsin, and later the Medical College of Wisconsin became aware of patients with vesiculo-pustular rash illnesses. North American prairie dogs were implicated as a potential exposure [14], and African rodents as a potential initial source of disease introduction to the U.S. Initially electron microscopy from prairie dog and human lesions demonstrated characteristic poxvirus particles. Coordination of efforts with the affected counties' and Wisconsin State Health Departments led to the hypothesis that this could be monkeypox disease [13]. Additional cases were reported from a neighboring state, Illinois, and then Indiana, with a variety of animal contact described including African, and North and South American rodents intended for the exotic pet trade. One of the initial case patients had maintained giant Gambian rats, previously found to be seroreactive against orthopoxviruses in a late 1990s monkeypox human outbreak in Africa. Samples were sent to CDC from human and later, animal cases. These samples tested positive for orthopoxvirus, and specifically, monkeypox in assays newly developed as part of the smallpox research agenda.

Research and response capacities developed as part of the smallpox research agenda were leveraged effectively to contain monkeypox disease spread in the U.S. in 2003 [19]. Education through outreach efforts and the LRN about the recognition of smallpox in 2001–2003, the availability of vaccine, and improved diagnostic and surveillance capacities as part of smallpox preparedness likely all contributed to the rapid response and containment of monkeypox disease. Issuance of a ban on the importation of African rodents and interstate movement of prairie dogs were additional critical strategies to prevent disease spread.

In turn, the U.S. outbreak provided additional insights into monkeypox disease. Although there were several patients with severe illness [14], there were no deaths in the 10 probable and 37 confirmed U.S. cases. Additional research led to the characterization of two clades of monkeypox (West African and Congo Basin) [8] and predicted protein differences that may be involved in altered pathogenicity. North American prairie dogs were found to be an effective species to transmit disease to one another, and to humans. The prairie dog also proved to be a model of systemic orthopoxvirus disease with similar parameters as that seen in human illness [5, 6] and has been successfully used to evaluate the antiviral effectiveness of a small molecule compound developed as part of the smallpox research agenda. Drug efficacy was shown not only in animals treated prophylactically or post exposure, but also in animals initiated on treatment well into the symptomatic phases of illness, after rash onset [16].

9.3 Ebola

Ebola was first identified as a human disease in 1976 in the Democratic Republic of Congo (DRC). Between 1976 and 2014, there were 35 reports of outbreaks of Ebola virus disease. The number of cases per outbreak was between 1 and 475. The 2014–2016 West African epidemic of >28,000 cases is 10 times greater than all cases of Ebola from prior outbreaks [1].

Ebola is a member of the filovirus family. The virus is a zoonotic infection of humans, and the reservoir is thought to be a species of fruit bat. In humans, fever or feeling feverish is one of the first signs of illness after a 2–21 day (8–10 days average) incubation period. Influenza-like symptoms including headache and myalgias manifest early. Later in the disease, abdominal pain, diarrhea, vomiting, bleeding, and hiccups are often noted. Transmission between humans fuels outbreaks and is via contact with infected excreta – blood, vomit, stool, or other excrement or secretions. Unsafe injection practices or inadequate personal protective equipment in handling patients in the healthcare setting can be a cause of nosocomial outbreaks. Mortality varies between 50% and 80% for different species of the virus.

Many factors may have contributed to the magnitude of the West African Ebola epidemic that claimed more than 11,000 lives. A weakened public health infrastructure undermined by years of civil unrest, a limited medical and public health workforce with minimal laboratory testing capacity, and transportation networks that permitted movement of individuals into urban areas in contrast to previous Ebola outbreaks in isolated, less-densely populated areas, were all components that likely contributed to sustained disease transmission. CDC was one of many international partners that participated in the response efforts to control disease spread. Ultimately, the outbreak was halted using classic public health tools including laboratory-based identification and isolation of cases, tracking of contacts of cases to rapidly isolate them, and in some cases quarantine if clinical illness manifested, improving infection control to better manage suspect and confirmed cases, and safe burial practices [1].

A coordinated and rapid response effort was critically important in controlling disease in a number of instances during the West African response. After an ill traveler from Liberia transited through the airport into Lagos, Nigeria, the outbreak was contained to three generations of disease, with 21 probable or confirmed cases. This was achieved by coordination of the emergency operations center established for polio eradication, and using persons in local epidemiology training programs, health officials, as well as international responders. Through their efforts almost 900 contacts were successfully followed, and an Ebola treatment unit was constructed on the grounds of an existing TB treatment facility to isolate and care for patients [1]. In Liberia, the Rapid Isolation and Treatment of Ebola (RITE) strategy was designed to improve responses to remote outbreaks of disease. Designated teams of trained responders went to areas where new cases were identified to provide interventions that included (1) engagement of traditional and community leaders in response activities; (2) community education about Ebola virus transmission and prevention; (3) active case finding, contact tracing and monitoring; (4) quarantine of asymptomatic high risk contacts at home or in designated quarantine facilities; (5) isolation and treatment of patients; and (6) safe burial practices. After the RITE strategy was introduced, the proportion of patients isolated and treated increased from a median of 28% to 81%, the median number of generations of disease fell from 4 to 2, and the case fatality fell from 87% to 50%. The size of outbreaks decreased from 10 to 64 cases to 4–28 cases, and the duration of outbreaks decreased from 52 to 90 days to 7–58 days [7].

The importance of polymerase chain reaction (PCR) laboratory capacity for nucleic acid testing was identified in prior outbreaks of Ebola virus disease in DRC and Uganda [18]] and was the focus of many international responders in West Africa. PCR assays were used to quickly identify cases, and in those with uncertain antecedent illness to confirm whether Ebola was the cause of death, to inform burial practices. Point-of-care lateral flow-based assays, developed during the outbreak, became available later in the response. Regulatory science reviews to understand assay performance characteristics provided the context for using appropriate diagnostics during the response with interim "WHO prequalified", or if developed in the U.S., emergency use authorization (EUA), approval for use.

In addition to diagnostics, many fundamental advances had been made in the development of antivirals and vaccines for Ebola virus disease, with none having completed regulatory scientific review. During the response to Ebola in West Africa, international efforts focused on rapid assessment and research evaluation of promising medical countermeasures. Likewise, in-country structures to permit these studies were largely developed during the outbreak. A summary of five antiviral studies, and four vaccine studies, are summarized in the 2017 IOM report [11]. To date, investigational therapeutics for Ebola virus disease have yet to demonstrate statistically significant survival benefit with respect to contemporary control arms. A trial of a VSV-vectored Ebola glycoprotein vaccine was completed, and prevented disease among contacts (and their contacts) of Ebola cases, and is reviewed in the prior reference. As well, the challenges and ethics of conducting human subjects clinical research on case patients and appropriate study designs during a response is reviewed. When undertaken, research needs to be integrated into the response to make sure the public health response is well coordinated to control disease transmission.

Near the end of the West African outbreak, molecular epidemiology data substantiated the transmission of Ebola via sexual contact [9]. These findings suggested the initiation of research studies to look at viral persistence. Studies have determined that Ebola virus RNA can persist longer than previously identified and as such, impacts guidelines for infection prevention and control recommendations for EVD survivors. In turn, informed public health prevention approaches using education and barrier sexual precautions should prevent additional spread of disease [17].

9.4 Summary

Research prepares us for response efforts, and provides the basis to develop and validate tools to recognize disease early, prevent spread of disease, and treat disease. We learn from response efforts, and this learning informs our next research efforts.

References

1. Bell BP, Damon IK, Jernigan DB, Kenyon TA, Nichol ST, O'Connor JP, Tappero JW (2016) Overview, control strategies, and lessons learned in the CDC response to the 2014–2016 Ebola epidemic. MMWR Suppl 65:4–11
2. Carroll D, Emerson G (2011) Chasing Jenner's VAccine. PLoS One 6(8):e23086
3. Damon I (2011) Poxviruses that infect humans. In: Murray PJ, Baron EJ, Jorgensen G, Landry M, Pfaller MA (eds) *Manual of clinical microbiology*. ASM Press, Washington, DC
4. Damon IK, Damaso CR, McFadden G (2014) Are we there yet? The smallpox research agenda using variola virus. PLoS Pathog 10:e1004108
5. Hutson CL, Carroll DS, Self J, Weiss S, Hughes CM, Braden Z, Olson VA, Smith SK, Karem KL, Regnery RL, Damon IK (2010) Dosage comparison of Congo Basin and West African strains of monkeypox virus using a prairie dog animal model of systemic orthopoxvirus disease. Virology 402:72–82
6. Hutson CL, Damon IK (2010) Monkeypox virus infections in small animal models for evaluation of anti-poxvirus agents. Viruses 2:2763–2776
7. Kateh F, Nagbe T, Kieta A, Barskey A, Gasasira AN, Driscoll A, Tucker A, Christie A, Karmo B, Scott C, Bowah C, Barradas D, Blackley D, Dweh E, Warren F, Mahoney F, Kassay G, Calvert GM, Castro G, Logan G, Appiah G, Kirking H, Koon H, Papowitz H, Walke H, Cole IB, Montgomery J, Neatherlin J, Tappero JW, Hagan JE, Forrester J, Woodring J, Mott J, Attfield K, DeCock K, Lindblade KA, Powell K, Yeoman K, Adams L, Broyles LN, Slutsker L, Larway L, Belcher L, Cooper L, Santos M, Westercamp M, Weinberg MP, Massoudi M, Dea M, Patel M, Hennessey M, Fomba M, Lubogo M, Maxwell N, Moonan P, Arzoaquoi S, Gee S, Zayzay S, Pillai S, Williams S, Zarecki SM, Yett S, James S, Grube S, Gupta S, Nelson T, Malibiche T, Frank W, Smith W, Nyenswah T (2015) Rapid response to Ebola outbreaks in remote areas – Liberia, July-November 2014. MMWR Morb Mortal Wkly Rep 64:188–192

8. Likos AM, Sammons SA, Olson VA, Frace AM, Li Y, Olsen-Rasmussen M, Davidson W, Galloway R, Khristova ML, Reynolds MG, Zhao H, Carroll DS, Curns A, Formenty P, Esposito JJ, Regnery RL, Damon IK (2005) A tale of two clades: monkeypox viruses. J Gen Virol 86:2661–2672
9. Mate SE, Kugelman JR, Nyenswah TG, Ladner JT, Wiley MR, Cordier-Lassalle T, Christie A, Schroth GP, Gross SM, Davies-Wayne GJ, Shinde SA, Murugan R, Sieh SB, Badio M, Fakoli L, Taweh F, de Wit E, van Doremalen N, Munster VJ, Pettitt J, Prieto K, Humrighouse BW, Stroher U, DiClaro JW, Hensley LE, Schoepp RJ, Safronetz D, Fair J, Kuhn JH, Blackley DJ, Laney AS, Williams DE, Lo T, Gasasira A, Nichol ST, Formenty P, Kateh FN, De Cock KM, Bolay F, Sanchez-Lockhart M, Palacios G (2015) Molecular evidence of sexual transmission of Ebola virus. N Engl J Med 373:2448–2454
10. McCollum AM, Damon IK (2014) Human monkeypox. Clin Infect Dis 58:260–267
11. National Academies of Sciences, Engineering, Medicine, Health, Division Medicine, Policy Board on Health Sciences, Health Board on Global, and Outbreak Committee on Clinical Trials During the – Ebola (2017) In: Busta ER, Mancher M, Cuff PA, McAdam K, Keusch G (eds) *Integrating clinical research into epidemic response: the Ebola experience*. National Academies Press, Washington, DC. Copyright 2017 by the National Academy of Sciences. All rights reserved
12. Petersen BW, Damon IK, Pertowski CA, Meaney-Delman D, Guarnizo JT, Beigi RH, Edwards KM, Fisher MC, Frey SE, Lynfield R, Willoughby RE (2015) Clinical guidance for smallpox vaccine use in a postevent vaccination program. MMWR Recomm Rep 64:1–26
13. Reed KD, Melski JW, Graham MB, Regnery RL, Sotir MJ, Wegner MV, Kazmierczak JJ, Stratman EJ, Li Y, Fairley JA, Swain GR, Olson VA, Sargent EK, Kehl SC, Frace MA, Kline R, Foldy SL, Davis JP, Damon IK (2004) The detection of monkeypox in humans in the Western Hemisphere. N Engl J Med 350:342–350
14. Reynolds MG, Davidson WB, Curns AT, Conover CS, Huhn G, Davis JP, Wegner M, Croft DR, Newman A, Obiesie NN, Hansen GR, Hays PL, Pontones P, Beard B, Teclaw R, Howell JF, Braden Z, Holman RC, Karem KL, Damon IK (2007) Spectrum of infection and risk factors for human monkeypox, United States, 2003. Emerg Infect Dis 13:1332–1339
15. Rubins KH, Hensley LE, Jahrling PB et al (2004) The host response to smallpox: analysis of the gene expression program in peripheral blood cells in a nonhuman primate model. Proc Natl Acad Sci U S A 101:15190–15195
16. Smith SK, Self J, Weiss S, Carroll D, Braden Z, Regnery RL, Davidson W, Jordan R, Hruby DE, Damon IK (2011) Effective antiviral treatment of systemic orthopoxvirus disease: ST-246 treatment of prairie dogs infected with monkeypox virus. J Virol 85:9176–9187
17. Soka MJ, Choi MJ, Baller A, White S, Rogers E, Purpura LJ, Mahmoud N, Wasunna C, Massaquoi M, Abad N, Kollie J, Dweh S, Bemah PK, Christie A, Ladele V, Subah OC, Pillai S, Mugisha M, Kpaka J, Kowalewski S, German E, Stenger M, Nichol S, Stroher U, Vanderende KE, Zarecki SM, Green HH, Bailey JA, Rollin P, Marston B, Nyenswah TG, Gasasira A, Knust B, Williams D (2016) Prevention of sexual transmission of Ebola in Liberia through a national semen testing and counselling programme for survivors: an analysis of Ebola virus RNA results and behavioural data. Lancet Glob Health 4:e736–e743
18. Towner JS, Rollin PE, Bausch DG, Sanchez A, Crary SM, Vincent M, Lee WF, Spiropoulou CF, Ksiazek TG, Lukwiya M, Kaducu F, Downing R, Nichol ST (2004) Rapid diagnosis of Ebola hemorrhagic fever by reverse transcription-PCR in an outbreak setting and assessment of patient viral load as a predictor of outcome. J Virol 78:4330–4341
19. Update: multistate outbreak of monkeypox–Illinois, Indiana, Kansas, Missouri, Ohio, and Wisconsin, 2003. (2003) MMWR Morb Mortal Wkly Rep 52:642–646

Chapter 10
Rapid and Low-Cost Tools Derived from Plants to Face Emerging/Re-emerging Infectious Diseases and Bioterrorism Agents

Rosella Franconi, Elena Illiano, Francesca Paolini, Silvia Massa, Aldo Venuti, and Olivia Costantina Demurtas

Abstract Whether naturally occurring or man-made, biological threats pose a severe risk in an increasingly globalized world.

The dual-use nature of biological research, with its most recent advances in biotechnology ('synthetic biology', gene editing, nanotechnologies etc.) and the rapid diffusion of knowledge, raise proliferation concerns of biological weapons by non-state actors.

Thus, there is an urgent need to develop measures intended to enhance diagnostic, prophylactic and therapeutic capabilities and capacities to improve the ability of society to combat infectious diseases outbreaks, as well as to alleviate the effects of bioterrorism attacks.

We present here two examples of biotechnology usage for biodefence purposes: (i) plants as biofactories for the rapid production of improved biopharmaceuticals ('Plant Molecular Farming'), and (ii) plant sequences as immune-modulating agents to enhance the efficacy of genetic vaccines.

These platforms represent two promising (and complementary) approaches for the rapid and low-cost production of countermeasures (diagnostics and vaccine candidates) against emerging, re-emerging and bioterrorism-related infections.

R. Franconi (✉) · E. Illiano · S. Massa · O. C. Demurtas
Department for Sustainability, Italian National Agency for New Technologies, Energy and the Environment (ENEA), 'Casaccia' Research Centre, Rome, Italy
e-mail: rosella.franconi@enea.it

F. Paolini · A. Venuti
HPV-UNIT – RIDAIT Department, UOSD Tumor Immunology and Immunotherapy, Regina Elena National Cancer Institute, Rome, Italy

V. Radosavljevic et al. (eds.), *Defence Against Bioterrorism*, NATO Science for Peace and Security Series A: Chemistry and Biology,
https://doi.org/10.1007/978-94-024-1263-5_10

Keywords Infectious disease · SARS-CoV · Biothreat · Bioweapon · Genetic vaccines · Molecular farming · Plant sequences · Diagnostics

10.1 Introduction

10.1.1 Biological Threats: Natural Infections and Biological Weapons

Infectious diseases represent a significant burden on public health and economic stability of societies all over the world being the cause of approximately 25% of 60 million of deaths (in developing countries this percentage reaches 45% of deaths).[1] In 2015, about 50% of all deaths among children under 5 years of age were due to infectious diseases.[2]

In a globalized world (more travels, trade and greater interconnectedness between countries) infectious disease outbreaks are becoming inevitable, and they remain unpredictable. When faced with diseases for which there are few or no medical countermeasures, massive chaos and considerable loss of lives can ensue: countries without adequate health services are the most vulnerable to the impacts of infectious diseases due to the difficulty in administering effective medical treatments at an early stage and putting effective preventative measures into place.

The spectrum of the biological risk represented by infectious diseases is continuous, including events than can be difficult to distinguish as natural (i.e. natural occurring pandemics, re-emerging infectious diseases, unintended consequences of research), accidental (due to laboratory accidents, ignorance or negligence) or intentional (due to sabotage or biowarfare). Whatever the origin is, '*a health threat anywhere is a health threat everywhere*'.[3] Thus, it is necessary to identify emerging epidemics as soon as possible, stop them before they spiral out of control and develop suitable medical countermeasures, such as novel and effective vaccines.

10.1.1.1 Natural Infections

Infectious diseases have been for centuries among the leading causes of death and changed the fate of entire civilizations [1]. 'Black Death' (plague) in the Middle Ages (1348–1350) killed 30–60% of Europe's population; smallpox, in the twentieth century, was responsible for 300–500 million deaths and decimated and

---

[1] World Health Organization, WHO, "The top 10 causes of death" January 2017 http://www.who.int/mediacentre/factsheets/fs310/en/index2.html).

[2] WHO, "Child mortality and causes of death", http://www.who.int/gho/child_health/mortality/causes/en/).

[3] https://publichealth.wustl.edu/a-health-threat-anywhere-is-a-health-threat-everywhere/

weakened native populations in the Americas and Australia in the eighteenth century, prior to its final eradication in the late 1970s; measles for centuries caused massive destruction to native populations especially in the Americas and Europe over the years: in 2000 it was declared eradicated in the US but, in spite of the availability of a safe and cost-effective vaccine, it continues to circulate in various parts of the world, causing many deaths globally particularly among young children.[4] Spanish Influenza epidemic (1918–1919) killed as many as 40 million people worldwide and it considered the most devastating epidemic recorded in world history and a global disaster [2].

Over the past decades, at least 30 novel infectious agents affecting humans have emerged, most of which are zoonotic and whose origins have been shown to correlate significantly with environmental/ecological (i.e. climate change, floods, change of agricultural practices, natural disasters, habitats destruction) and socioeconomic (i.e. increase in population density, falling living standards, decline of infrastructures, human travels, conflicts and social instability, killing of wild animals for food) factors. These factors, together with the natural evolution of pathogens, are constantly leading to facilitate infections in humans, changing the nature of biological risks and increasing the global impact [3]. In particular, *newly emerging* infections refer to diseases that have been discovered in the human host for the first time (i.e. the severe acute respiratory syndrome –SARS- coronavirus, SARS-CoV) while *reemerging infectious diseases* can be defined as infectious diseases that reappear, usually in more pathogenic form and in rapidly increasing incidence or new geographic locations after apparent control or eradication (i.e. Filoviruses like Ebola and Marburg).[5]

The field of emerging disease exploration has been strengthened by the creation of dedicated emerging diseases units and programmes at the Centre for Disease Control and Prevention (CDC)[6] or at the European Centre for Disease Prevention and Control (ECDC).[7] These institutions monitor current infectious disease outbreaks, assess the risk to public health and provide technical support to the US/EU level-response to such threats.

10.1.1.2 Bioweapons

The emergence of some pathogens can be the result of deliberate human action, being employed as biological weapons (or 'bioweapons', 'biological warfare', BW) for destruction. Among the so-called 'CBRN'[8] weapons, biological weapons include

[4] WHO, "Measles," http://www.who.int/mediacentre/factsheets/fs286/en/

[5] WHO, "Emerging diseases", http://www.who.int/topics/emerging_diseases/en/

[6] WHO, "Global infectious disease surveillance," http://www.who.int/mediacentre/factsheets/fs200/en/

[7] https://ecdc.europa.eu/en/about-us/who-we-are/disease-programmes/emerging-and-vector-borne-diseases-programme

[8] CBRN: chemical, biological, radiological and nuclear.

deadly pathogens – bacteria or viruses – or toxins that can be deliberately released in order to cause harm to people or animals and plants ('agroterrorism'). In addition to potentially catastrophic immediate impact, these agents could also trigger long-term disasters, causing regional instability and challenging international security.[9]

Biological agents can be easily grown and disseminated through inhalation, ingestion or skin absorption. Some of them might affect large numbers of people (such as the highly contagious SARS-CoV), while others might be less contagious but more deadly for those they affect (such as Ebola). Since bioweapons use could resemble natural pandemics, it would be very difficult to differentiate between naturally occurring infections and those resulting from malicious use.

In spite of the difficulties in the evaluation of BW true frequency of use and impact in the past, (due to lack of data, manipulation/secret by political authorities etc.), historical analysis has shown that biological agents have been used in several occasions from ancient times through to the twenty-first century to cause panic and terror among civil populations (for a comprehensive history of biological warfare see Barras and Greub [4]).

The Biological Weapons Convention (BWC, signed in 1972 and entered into force in 1975)[10] bestows a prohibition on the weaponisation of biological pathogens and agents.[11] In particular, BWC prohibits: (i) the possession of biological agents except for 'prophylactic, protective, or other peaceful purposes'; (ii) the development of technologies intended for the dispersal of biological agents for offensive military purposes; and (iii) the destruction of existing stocks.[12]

Despite the destructive potential of bioweapons and the relative ease with which malicious actors could obtain many of the materials and know-how required to build them, relatively few cases of bioterrorism or sabotage have been recorded in the twentieth and twenty-first centuries: in the period 1970–2014, of a global total of 143 CBRN attacks, 35 used BW.[13] A potential disincentive for the acquisition and use of BW might be represented by the fact that biological agents are indiscriminate, and cannot be easily contained once released. On the other hand, some specialists have raised the question whether bioterrorism is a myth or reality [5]

[9] Use of Chemical, Biological, Radiological and Nuclear Weapons by Non-State Actors. Emerging trends and risk factors. Lloyd's Emerging Risk Report – 2016, Chatman House, The Royal Institute of International Affairs, p 31. https://www.lloyds.com/news-and-insight/risk-insight/library/society-and-security/cbrn

[10] The BWC currently has 178 states-parties and six signatories (Central African Republic, Egypt, Haiti, Somalia, Syria, and Tanzania). Twelve states have neither signed nor ratified the BWC (Chad, Comoros, Djibouti, Eritrea, Israel, Kiribati, Micronesia, Namibia, Niue, Samoa, South Sudan and Tuvalu).

[11] An analogous Chemical Weapons Convention (CWC, signed in 1993 and entered into force in 1997) incorporates a general clause prohibiting the weaponisation of all chemicals. After dedicated UN negotiates, the Treaty on the Prohibition of Nuclear Weapons (TPNW) has been signed by 122 countries only recently (July 7th 2017, http://undocs.org/A/CONF.229/2017/8).

[12] https://www.armscontrol.org/factsheets/bwcsig

[13] National Consortium for the Study of Terrorism and Responses to Terrorism (START), 2015. Global Terrorism Database [online]. Available at: http://www.start.umd.edu/gtd/

while some others consider BW as a '*common aspect of the human behavioural repertoire*' [6].

As knowledge diffuses rapidly to different parts of the world, through the globalisation of information and communications technology, a growing concern emerges that biologic agents could be easily used as weapons by non-state actors (such as terrorist organisations, saboteurs or lone actors) in the future. Moreover, certain emerging technologies and scientific advances of biotechnology (i.e. nanotechnology, synthetic biology, gene editing) are altering the risk landscape for bioweapons use in a variety of ways ('*dual use*' concept). The malicious use of synthetic or edited pathogens could possibly enable hostile actors to develop weapons that are cheaper, more powerful and easier to use (i.e. deadly viruses such as polio and Ebola can be synthesized using public databases and available technology).

Being highly unlikely that societies could ever completely eliminate vulnerability to biological agents, there are no doubts that the topic of biological warfare poses very difficult problems, opening some novel challenges in the ethical domain [7].

10.1.1.3 Countermeasures and Novel Approaches to face Biothreats

Since biological agents might be more lethal than chemical weapons, more difficult to detect than nuclear weapons and less expensive to be produced using the common technologies available in any biological laboratory, measures intended to enhance mitigation (diagnostic, surveillance, etc.) and adaptation (new therapeutics, vaccines, etc.) capabilities and capacities, alongside with training and education, will improve the ability of society to combat 'regular' infectious diseases outbreaks, as well as counteracting the effects of bioterrorist attacks, enhancing society's resilience.

Novel technologies (such as nanotechnology, new detection technologies, next generation sequencing) could be useful for clean-up and detection, preserving the health and well-being of first responders or assisting local law enforcement in identifying the nature of an outbreak/attack and the kinds of biological agents involved, making responding easier, reducing the destructive and disruptive capacity of biological threats. On the other hands, antimicrobials and vaccines offer possible means for protection toward sudden emerging infectious diseases outbreaks, both natural and intentional. For these, it is important to rely on strategic reserves of therapeutics/vaccines against known biothreat agents as well as having tools/platforms for the rapid production of effective countermeasures against (novel) pathogens.

In general, the pharmaceutical industry is not much involved in vaccines and rarely invests in research and development for diseases with limited market incentive: when the Ebola outbreak began in 2014, vaccine candidates were unavailable because they had stalled in the pipeline. Repeated outbreaks (most recently, Ebola and Zika) have forged a global consensus that current models for developing vaccines for sporadic epidemic are not working, and that a new system is urgently needed, also in the light that '*Pathogens are not only terrifying, they're expensive*'[14] (the 2003 SARS epidemic cost $30 billion in only 4 months). Thus, novel global approaches are needed to drive product innovation to prevent and contain future infectious diseases epidemics.

A few years ago, the Defense Advanced Research Projects Agency (DARPA, US) started supporting new technologies that radically accelerate the manufacturing of protein vaccines and protein-based therapeutics. In 2007, after realizing that low-cost, plant-derived vaccines are better tools to control many infectious diseases in humans, DARPA financed projects for the development of cGMP facilities for plant-made vaccines. In 2015, DARPA funded Inovio Pharmaceuticals, Inc. with $45 million to develop multiple treatment and prevention approaches against Ebola (a DNA-based vaccine against Ebola, a therapeutic DNA-based monoclonal antibody product and a conventional monoclonal antibody to treat Ebola). More recently, the Biological Technologies Office (BTO) of DARPA sponsored the 'Pandemic Prevention Platform' (P3)[15] program whose goal is to achieve an integrated capability that can deliver pandemic prevention countermeasures to patients within 60 days of an outbreak, changing outbreak response by enabling rapid discovery, characterization, production, and testing of efficacious medical countermeasures (i.e. generation of virus stock, including viral unknowns; rapid evolution of antibody candidates; gene-encoded antibody delivery methods).

Another initiative is represented by the Global Health Security Agenda[16] (GHSA), launched in February 2014, a growing partnership of over 50 nations, international organizations, and non-governmental stakeholders. It pursues a multilateral and multi-sectoral approach to strengthen both the global capacity and nations' capacity to prevent, detect, and respond to infectious diseases threats whether naturally occurring, deliberate, or accidental. The idea is that, this capacity, once established, would mitigate the devastating effects of threats posed by highly pathogenic infectious diseases and bioterrorism events by rapidly detecting and transparently reporting outbreaks when they occur, and employing an interconnected global network that can respond effectively to limit the spread of infectious disease outbreaks in humans and animals, mitigate human suffering and the loss of human life, and reduce economic impact.

Besides the so-called "One Health Initiative" (strategies to control diseases across species),[17] more recently (January 2017) the Coalition for Epidemic Preparedness Innovations (CEPI)[18] was launched at the World Economic Forum. It is a partnership of public, private, philanthropic and civil organizations to accelerate (safe and affordable) vaccine development for emerging infectious diseases,

---

[14] https://pandorareport.org/2017/06/30/pandora-report-6-30-2017/

[15] https://globalbiodefense.com/2017/04/18/dstl-darpa-intercept-evolving-countermeasures-bioterrorism/

[16] https://www.ghsagenda.org/

[17] http://www.onehealthinitiative.com/

[18] http://www.who.int/medicines/ebola-treatment/TheCoalitionEpidemicPreparednessInnovations-an-overview.pdf

particularly for diseases that lack market incentives, readying pandemic defences during peacetime. It is based on a memorandum of understanding with the World Health Organization, (WHO), and has established a partnership with the Bill & Melinda Gates Foundation, governments (like India, Germany, Japan, Norway), industry partners and private funders (i.e. Wellcome Trust), academic institutions and civil society organisations among others. According to the WHO 'R&D Blueprint for Action to Prevent Epidemics' (that indicates the priority pathogens against which the development of medical countermeasures are urgently needed)[19] and based on specific criteria (such as risk of an outbreak occurring, transmissibility of the pathogen, burden of disease, feasibility of vaccine development and the current pipeline candidates), as a first step, three diseases were selected (Lassa fever, Nipah virus, and Middle East respiratory syndrome coronavirus, MERS-CoV) to move new vaccines from preclinical to proof of principle studies in humans. However, since there always will be an unknown or a not selected pathogen that it will not be possible to predict, CEPI aims also to support the development of rapid and adaptable vaccine technology platforms, where antigens from a new pathogen can substitute or be added to an existing vaccine.

10.2 Novel Platforms for Vaccine Production

Traditional vaccines against infectious diseases are failing to satisfy the global demand because of limited scalability of production systems and long production timelines (similar issues are applicable to other anti-infective agents). This is especially a problem for emerging pathogens that carry the inherent risk of pandemic spread in a naïve population.

A considerable number of different platform technologies are under development and, among these, plant-derived vaccines and plasmid-based DNA vaccines are encouraging tools.

Plants have emerged as promising platforms for the production of subunit vaccines, monoclonal antibodies and other recombinant therapeutic proteins ('Plant Molecular Farming') due to time and cost efficiency, scalability, lack of harboured mammalian pathogens and ability to perform eukaryotic post-translational protein modification.

The recent apparent success in fighting Ebola virus with plant-made human antibodies put a spotlight on the enormous potential of this platform for applications in human health [8]. So far, several candidate countermeasures against emerging, re-emerging and bioterrorism-related infections have been produced in plants (reviewed in Rybicki [9]; Streatfield et al. [10]). The modularity of molecular engineering provides fast and scalable systems to be used in response to new outbreaks of highly infectious diseases with pandemic potential, such as influenza, malaria, and SARS [11].

[19] http://www.who.int/blueprint/priority-diseases/en/

In addition, genetic vaccines represent another advantageous platform for the rapid development of novel vaccines to face deliberate or naturally occurring outbreaks due to ease of preparation and general stability at room temperature. In this case plants might be exploited as a source of immune-modulating sequences able to increase the 'visibility' to the immune system of weak antigens for the construction of more powerful genetic vaccines.

10.2.1 Plants as Biofactories for the Production of Biopharmaceuticals ('Plant Molecular Farming', PMF)

Herbal medicine has formed the basis of health care throughout worldwide since the dawn of civilization, having been extensively utilised by ancient civilisations [12]. Thousands of plant species contributed to the development of important therapeutic drugs used in modern medicine: almost 50% of the synthetic medicines derive from phytochemicals and almost 30% of all pharmaceuticals approved by the US Food and Drug Administration (FDA) have a botanical origin (digoxin, morphine, salbutamol and aspirin represent some successful examples).

The use of plants as bioreactors ('Plant Molecular Farming', PMF) is a relatively new bioscience.[20] So far, a variety of subunit vaccines, monoclonal antibodies and therapeutic proteins have been produced in plants and other 'green' systems [13, 14] including candidate countermeasures against emerging, re-emerging and bioterrorism-related infections [10].

Plants represent ideal platforms for recombinant protein production for several reasons. Lower manufacturing costs have been widely assumed as an intrinsic advantage of plant-based production platforms. Biologic production in plants does not require capital-prohibitive facilities, bioreactors, and expensive culture media but can be easily scaled in relatively inexpensive greenhouses with simple mineral solutions. Plants compete with other expression systems for reduced risks of contamination with human/animal pathogens and ability to perform eukaryotic post-translational protein modification, such as glycosylation. There are differences in N-glycan and O-glycan structures between plants and mammals [15, 16]; nevertheless, the possibility to control the glycosylation pattern ('glyco-engineering' or plant 'glyco-biotechnology') provides a method for producing proteins with unique and uniform mammalian post-translational modifications, resulting in biologics with increased efficacy with respect to their mammalian cell-produced counterparts ('bio-betters') [17].

---

[20] It officially entered in the plant science field in the year 2000 as a specific session at the 6th Congress in Plant Molecular Biology, Quebec, Canada.

Recombinant proteins can be selectively expressed in particular plant cell compartments (chloroplast, apoplast etc.) or organs (seeds, roots, tubers etc.), where they are more stable and do not interfere with vegetative growth. Plant cells/tissues/organs can be lyophilized and stored at ambient temperature for many years, maintaining activity of expressed protein drugs. A promising approach is the use of edible plant tissues/organs expressing biopharmaceuticals for direct oral delivery, with no need for exhaustive purification, thus eliminating expensive downstream purification, cold storage and transportation costs [18]. This could be particularly useful for veterinary vaccines against major zoonotic diseases [19].

Furthermore, plant-based expression platforms offer safe, inexpensive and potentially limitless ways to produce therapeutics in a quick and flexible manner. If time and expression level might represent a limit of the transgenic technology, it can be overcome by transient expression mediated by plant viruses or by agroinfiltration [20]. Recently, novel transient expression vectors have been developed that allow the production of vaccines and therapeutics at unprecedented speed [21]. The recent apparent success[21] in fighting Ebola outbreak of 2014–2016 with a plant-made drug (ZMapp™, a cocktail of three human monoclonal antibodies) brought renewed attention to the field of plant-made biologics for human health whose potential and capacity to produce 'rapid response' vaccines had been already demonstrated by the commitment of several US companies in the production of 100 million doses of influenza vaccine a month by using such technology [9].

Another field in which plants could represent ideal production systems is that of antigen preparation for the development of diagnostic test that is particularly useful when a pathogen cannot be grown in the lab or is highly virulent and needs a methodology for safe, fast and affordable production or when it is necessary to rely on 'high quality' reagents.

Our early efforts in the field of PMF were focused in the expression of intracellular antibodies ('intrabodies') to obtain plants resistant to viral infection ('plantibody'-mediated resistance). In particular, we dealt with the Cucumber Mosaic Virus, CMV [22] and the Tomato Spotted Wilt Virus (TSWV, family *Bunyaviridae*) that, since the introduction of the vector *Frankliniella occidentalis* in Europe, become one of the limiting factors and one of the most serious threats to vegetable crops in the Mediterranean basin [23].

Later on, we focused on the use of plant-based platforms for the production of recombinant proteins for the development of novel protection/therapy tools and diagnostics to be quickly manufactured, at low cost and with minimal risk against infective agents like the human papillomavirus (HPV) [24–28] or the severe acute respiratory syndrome (SARS) coronavirus, SARS-CoV [29].

[21] The conditions of two infected American health aid workers dramatically improved soon after receiving the plant-derived experimental drug.

10.2.1.1 Case Study 1: Plant Derived SARS-CoV Antigens as Tools for Preventive Vaccines and Diagnostics

Severe acute respiratory syndrome (SARS) emerged in 2002, spreading to 29 countries over 5 continents, leading to more than 8000 infected patients globally[22] with a fatality rate of 9.6%. The aetiological agent of the syndrome, rapidly identified as a coronavirus (SARS-CoV), crossed the species barrier to infect humans, showing high morbidity and mortality rates. The end of the SARS outbreak was declared by WHO in July 2003. However, several local outbreaks were subsequently reported in China as a consequence of accidental laboratory contaminations or infections after contact with animals infected with SARS-CoV strains significantly different from those predominating in the 2002–2003 outbreak [30].

For its high transmissibility, high lethality and significant impact on the public health system, SARS-CoV has been defined a class C biological weapon (Centers for Disease Control and Prevention [31]). Currently, there are no approved antiviral treatments for SARS-CoV. Since a SARS epidemic may recur at any time in the future, it has been included in the WHO 'R&D Blueprint for Action to Prevent Epidemics' list and multiple therapeutic approaches against SARS-CoV (and MERS-CoV) are currently under development [32]. A recent example of such efforts is represented by the nucleotide prodrug GS-5734 (currently in clinical development for treatment of Ebola virus disease) that showed inhibition of SARS-CoV and MERS-CoV replication in multiple *in vitro* systems and in a mouse model of SARS-CoV pathogenesis [33]. Another important key to prevent and control a future outbreak of SARS is to develop novel rapid and specific diagnostic methods, in addition to those already available[23] so that suspected patients can be correctly triaged and isolated.

SARS-CoV has four major viral structural components, the spike (S), envelope (E), membrane (M) and nucleocapsid (N) proteins and 16 non-structural proteins [34].

The structural N and M proteins are the most abundant proteins, respectively, in the virus core and in the viral envelope. The N protein, expressed at early stages of infection, triggers an early, powerful antibody response by the host, thus it is considered the best diagnostic target [35]. Furthermore, since the N protein is able to induce a long-term cell-mediated immune response in animal models, it represents a potential vaccine candidate. The production of recombinant N protein has been achieved in a variety of heterologous expression systems, with eukaryotic platforms (such as insect cells, yeast) allowing more efficient and specific diagnostic tests [36].

The membrane M glycoprotein is functionally involved in the assembly and budding of virions from the cell. The M protein contains T cell epitopes [37] and the availability of recombinant M protein, in combination with other recombinant viral

---

[22] WHO Library Cataloguing in Publication Data (2006). SARS: how a global epidemic was stopped ISBN 92 9061 213 4.

[23] WHO, Severe Acute Respiratory Syndrome (SARS): Laboratory diagnostic tests, 2003) http://www.who.int/csr/sars/diagnostictests/en/

proteins might overcome the concern about the sensitivity and the specificity of N protein-based assay [38–40], thus providing high quality reagents to detect antibodies in the infected human host.

Recently, we demonstrated that plant transient expression systems can be used to produce SARS-CoV N and M antigens [29]. The N and M full-length genes, derived from the Frankfurt I isolate of human SARS-CoV [41], were inserted into different plant expression vectors.

The N protein was expressed in *Nicotiana benthamiana* plants using Potato Virus X (PVX)-mediated infection. The protein was obtained in systemic leaves of 100% infected plants. Differently from the N protein produced in bacteria, the plant-produced N protein doesn't display any proteolysis, demonstrating the suitability of the plant platform for the production of recombinant SARS-CoV antigens (Fig. 10.1).

In addition, we demonstrated that both crude extracts containing N protein, or purified plant-produced N protein, were specifically recognized in immunoblotting by sera derived from Chinese SARS convalescent patients of the 2003 outbreak, and not from patients affected by unrelated respiratory diseases. This study represents the first demonstration that the plant-derived N protein is able to reveal, by direct serology, human N-specific antibodies present in sera of SARS patients, thus pro-

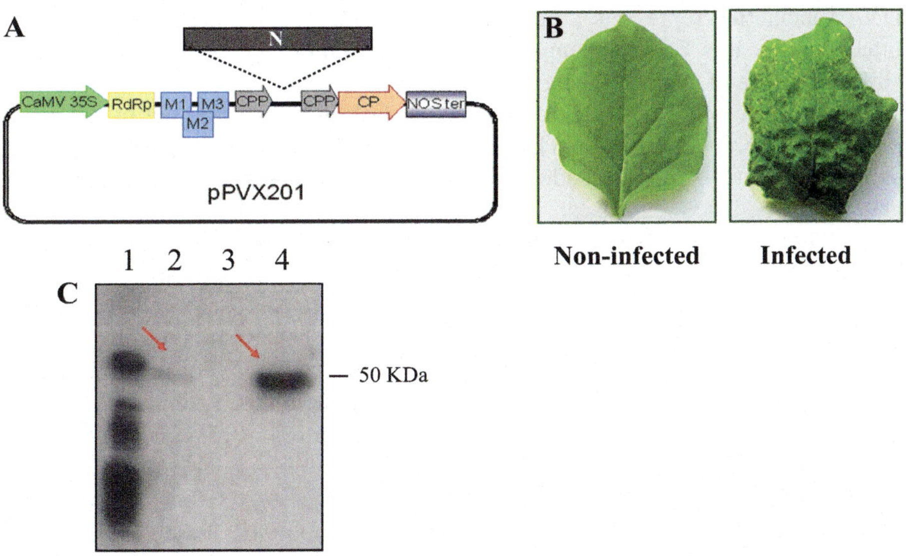

Fig. 10.1 Transient expression of the SARS-CoV N antigen in *Nicotiana benthamiana* leaves. (**a**) The *N* gene was inserted into a Potato virus X-derived vector (pPVX201). (**b**) The obtained construct was used to infect *N. benthamiana* plants. Infection spread systemically from inoculated leaves to apical leaves, where typical PVX-infection symptoms appeared 7–10 days post-inoculation. (**c**) Immunoblotting performed with a specific anti-N antibody showed the presence of the N protein (50 kDa) in both inoculated (lane 2) and symptomatic apical leaves (lane 4), but not in non-infected leaves (lane 3). The N protein produced in bacteria (lane 1) presents several degradation products (extra bands)

viding an adequate instrument to develop a rapid, low-cost, immune-based diagnostic assay to be used as an alternative or in association to molecular diagnosis.

For the M protein, we demonstrated that the wild type protein is toxic when expressed in bacteria and only a mutated form was obtained in this system, accumulating in the inclusion bodies. On the contrary, we demonstrated that plants allowed the expression of the full-length original M protein. In particular, we obtained a soluble M protein in *N. benthamiana* plants, using *Agrobacterium tumefaciens*-mediated infection. The reduced electrophoretic mobility observed for the plant-derived M protein, compared to that produced in bacteria, suggests the presence of glycosylation (the native M protein is N-glycosylated at the fourth residue), provided by this eukaryotic system.

These results provide a proof of principle for using plants as a robust, rapid and flexible production system for protein reagents suitable to face potential recurring SARS-CoV outbreaks.

10.2.2 Improved Genetic Vaccines Including Plant Immune-Modulating Sequences

DNA vaccination represents a new milestone in the technological efforts against infectious diseases, offering many advantages over other vaccine approaches due to simplicity, ease of manufacturing and safety.

DNA vaccines are currently used in veterinary medicine but one of the main problems to be solved for human DNA vaccines (both preventive and therapeutic) is their poor ability to induce an adequate immune response (production of antibodies and/or cell-mediated responses).

Several strategies have been developed to improve DNA vaccine efficacy (i.e. codon optimization, transfection reagents, roots of administration, adjuvants, combination with heterologous boosts). Increased understanding of molecular events driving innate and adaptive immune responses has assisted development of molecular adjuvants for DNA vaccine use. Such adjuvants comprise plasmid-encoded signalling molecules including cytokines, chemokines, immune costimulatory molecules, toll-like receptor agonists or inhibitors of immune suppressive pathways [42].

Another possibility of DNA immunization in combatting the threat of emerging infectious diseases, is to offer a unique and powerful approach to the production of high-quality antibodies (polyclonal, monoclonal or recombinant antibodies from phage display libraries) against various pathogens [43]. Compared with traditional protein-based immunization approaches, DNA immunization is efficient for testing novel immunogen design, does not require the production or purification of proteins from a pathogen or the use of recombinant protein technology and is effective at generating antibodies against conformation-sensitive targets.

Recent clinical data have shown that novel DNA vaccines design are able to induce high-level antigen-specific antibody responses [44] but the search of innovative immune-stimulatory sequences with few clinical use constraints (i.e., possible auto-immune responses induced by proteins of human origin) is still an open field.

In the following, examples of successful use of sequences of plant origin as immune-enhancers with the purpose of reinventing vaccine design are reported.

10.2.2.1 Case Study 2: Plant Derived Sequences for Improved Genetic Vaccines Against Infectious Agents

Several years ago we demonstrated that therapeutic (anti-cancer) DNA vaccines can be potentiated by using plant immune-modulating sequences. The driving idea was that some plant proteins, involved in plant defence responses (due to some similarities between mammalian and plant immune mechanisms) might have effects on tumours and human immunity through modulation of innate immune functions. This turned out to be possibly true and induced a tumour-Specific Antigen (TSA)-linked adaptive cell-mediated immunity (crucial for cancer resolution) [45, 46].

Recently, we developed a genetic vaccine where a plant protein signal sequence (ss-), was fused to the N-terminal portion of crucial viral antigens derived from the human papillomavirus type 16, HPV 16 (synthetic/fusion genes derived from E7 and L2 proteins) [47, 48].

Mammalian cells (HEK-293) were transfected to study the transient expression and the intracellular fate of the proteins encoded by the novel DNA constructs. In the case of a ss-E7 construct, the protein was found in the culture medium of transfected cells, whereas E7 without ss- was only present in the cell lysates, demonstrating the ability of the plant signal sequence to modulate the sorting of a heterologous protein in mammalian cells. The plant ss- was found to modify the processing also of other constructs (i.e. ss-E7-CP, where the E7 gene is fused to the coat protein of potato virus X), even though secretion was not observed in the culture medium, while for ss-L2 the protein was detected mostly into the cytoplasm of transfected cells.

The immunological effects of the ss- provided DNA vaccines were studied in animal models for HPV (C57BL/6 mice) with a prime/boost schedule, implying the use of electroporation (EP) after intra-muscular immunization, demonstrating that the plant signal sequence enhances the humoral response to DNA-based vaccines.

Electroporation (EP), indeed, appears a promising approach for improving immunogenicity of DNA vaccines for its ability to increase cellular permeability resulting in a high level of protein expression and improved immune response, as recent clinical trials have shown [49, 50]. A vaccination schedule, comprising a 'prime' with DNA plasmid at time zero and 'boost' with DNA at 1-week interval by intradermal (ID) + EP immunization, resulted in the induction of a strong humoral immune response, confirming that ID + EP in more efficient than intra-muscular (IM) vaccination. In particular, the ss-$L2_{1\text{-}200}$-E7 plasmid was able to produce the highest titers of both anti-$L2_{1\text{–}200}$ and anti-E7 IgGs. The EP immunization protocol determined also a longstanding humoral immune response against $L_{21\text{-}200}$, persisting, at least, 6 months in the utilized mouse model. Preliminary experiments seem to indicate the neutralizing nature of the anti-L2 antibodies.

To our best knowledge, this is the first demonstration that a signal sequence of a plant protein may exert a biological activity in mammalian cells and enhance immunogenicity of an antigen of interest. This approach might work also for other antigens (even if relatively large or glycoproteins) and for different pathogens, opening new perspectives in the design of DNA vaccines, especially to counteract infections where a fast and effective humoral response is needed. Such genetic vaccines can be easily produced on an industrial scale according to GMP, providing more effective and safe vaccines that do not involve the production of chemo/cytokines which might induce secondary responses, or of animal antigens that could cause cross autoimmune responses.

10.3 Conclusions and Perspectives

In order to avoid the devastating loss of life by (possible) viral outbreaks such as a next Ebola, Zika, avian flu, MERS [51] or a biological warfare (BW) attack, whose epidemiology is associated to sudden and unforeseen contagious burst, it is necessary to rely on small stockpiles ready when the next outbreak begins. At the same time it is fundamental to invest in technical platforms able to cut down the time to tailor the eventual vaccine candidate to be effective to the epidemic. In other words, when outbreaks happen, the vaccines will be ready in just few weeks/months for field-testing and mass-manufacture.

The two platforms we introduced, plants as bioreactors for the production of biopharmaceuticals ('Plant Molecular Farming'), and improved genetic vaccines endowed of plant sequences with immune-modulating activity, represent two promising (and complementary) approaches for the rapid and affordable production of countermeasures (diagnostics and vaccine candidates) against emerging, re-emerging and bioterrorism-related infections.

References

1. Hays JN (2005) Epidemics and pandemics: their impacts on human history. ABC-CLIO, Santa Barbara
2. Shanks GD, Brundage GF (2012) Pathogenic responses among young adults during the 1918 influenza pandemic. Emerg Infect Dis 18:201–207
3. Nii-Trebi NI (2017) Emerging and neglected infectious diseases: insights, advances, and challenges. BioMed Res Int 2017:5245021. https://doi.org/10.1155/2017/5245021. 15 p
4. Barras V, Greub G (2014) History of biological warfare and bioterrorism. Clin Microbiol Infect 20(6):497–502. https://doi.org/10.1111/1469-0691.12706

5. Mukunda G, Oye KA, Mohr SC (2009) What rough beast? Synthetic biology, uncertainty, and the future biosecurity. Polit Life Sci 28(2):2–26
6. Rózsa L (2009) The motivation for biological aggression is an inherent and common aspect of the human behavioural repertoire. Med Hypotheses 72:217–219
7. Lederberg J (1999) Introduction. In: Lederberg J (ed) Biological weapons. Limiting the threat. MIT, Cambridge, MA, pp 3–5
8. Arntzen C (2015) Plant-made pharmaceuticals: from 'Edible Vaccines' to Ebola therapeutics. Plant Biotechnol J 13(8):1013–1016
9. Rybicki EP (2014) Plant-based vaccines against viruses. Virol J 11:205. https://doi.org/10.1186/s12985-014-0205-0
10. Streatfield SJ, Kushnir N, Yusibov V (2015) Plant-produced candidate countermeasures against emerging and reemerging infections and bioterror agents. Plant Biotechnol J 13(8):1136–1159. https://doi.org/10.1111/pbi.12475
11. De Martinis D, Rybicki EP, Fujiyama K et al (2016) Editorial: plant molecular farming: fast, scalable, cheap, sustainable. Front Plant Sci 7:1148. https://doi.org/10.3389/fpls.2016.01148
12. Petrovska BB (2012) Historical review of medicinal plants' usage. Pharmacogn Rev 6(11):1–5
13. Franconi R, Demurtas OC, Massa S (2010) Plant-derived vaccines and other therapeutics produced in contained systems. Expert Rev Vaccines 9(8):887–892
14. Yao J, Weng Y, Dickey A, Wang KY (2015) Plants as factories for human pharmaceuticals: applications and challenges. Int J Mol Sci 16(12):28549–28565. https://doi.org/10.3390/ijms161226122
15. Lannoo N, Van Damme EJ (2015) Review/N-glycans: the making of a varied toolbox. Plant Sci 239:67–83. https://doi.org/10.1016/j.plantsci.2015.06.023
16. Strasser R (2016) Plant protein glycosylation. Glycobiology 26(9):926–939
17. Schoberer J, Strasser R (2017) Plant glyco-biotechnology. Semin Cell Dev Biol. pii: S1084-9521(16)30360-3. https://doi.org/10.1016/j.semcdb.2017.07.005
18. Sohrab SS, Suhail M, Kamal MA et al (2017) Edible vaccine: current status and future perspectives. Curr Drug Metab. https://doi.org/10.2174/1389200218666170711121810
19. Shahid N, Daniell H (2016) Plant-based oral vaccines against zoonotic and non-zoonotic diseases. Plant Biotechnol J 14(11):2079–2099. https://doi.org/10.1111/pbi.12604
20. Gleba Y, Klimyuk V, Marillonet S (2007) Viral vectors for the expression of proteins in plants. Curr Opin Biotech 18:134–241. https://doi.org/10.1016/j.biotechadv.2010.01.005
21. Kopertekh L, Schiemann J (2017) Transient production of recombinant pharmaceutical proteins in plants: evolution and perspectives. Curr Med Chem. https://doi.org/10.2174/0929867324666170718114724
22. Villani ME, Roggero P, Bitti O et al (2005) Immunomodulation of a plant virus infection by intrabodies selected in vitro from a stable single-framework phage display library. Plant Mol Biol 58:305–316
23. Franconi R, Roggero P, Pirazzi P et al (1999) Functional expression in bacteria and in plants of an scFv antibody fragment against tospoviruses. Immunotechnology 4(3–4):189–201
24. Franconi R, Venuti A (2006) HPV vaccines in plants: an appetising solution to control infection and associated cancers. In: Campo MS (ed) Papillomavirus research: from natural history to vaccines and beyond. Caister Academic Press, Norfolk, pp 357–372
25. Franconi R, Di Bonito P, Dibello F et al (2002) Plant-derived Human Papillomavirus 16 E7 oncoprotein induces immune response and specific tumor protection. Cancer Res 62:3654–3658
26. Giorgi C, Franconi R, Rybicki EP (2010) HPV vaccines in plant. Expert Rev Vaccines 9(8):913–924
27. Massa S, Franconi R, Brandi R et al (2007) Anti-cancer activity of plant-produced HPV16 E7 vaccine. Vaccine 25:3018–3021
28. Venuti A, Massa S, Mett V et al (2009) An E7-based therapeutic vaccine protects mice against HPV16 associated cancer. Vaccine 27:3395–3397
29. Demurtas OC, Massa S, Illiano E et al (2016) Antigen production in plant to tackle infectious diseases flare up: the case of SARS. Front Plant Sci 7:54. https://doi.org/10.3389/fpls.2016.00054

30. Peiris JS, Guan Y, Yuen KY (2004) Severe acute respiratory syndrome. Nat Med 10(12 Suppl):S88–S97. https://doi.org/10.1038/nm1143
31. Centers for Disease Control and Prevention (CDC), Department of Health and Human Services (HHS) (2012) Possession, use, and transfer of select agents and toxins; biennial review. Final rule Fed Regist 77(194):61083–61115
32. Zumla A, Chan JF, Azhar EI et al (2016) Coronaviruses – drug discovery and therapeutic options. Nat Rev Drug Discov 15(5):327–347. https://doi.org/10.1038/nrd.2015.37
33. Sheahan TP, Sims AC, Graham RL et al (2017) Broad-spectrum antiviral GS-5734 inhibits both epidemic and zoonotic coronaviruses. Sci Transl Med 9(396)
34. Bartlam M, Xu Y, Rao Z (2007) Structural proteomics of the SARS coronavirus: a model response to emerging infectious diseases. J Struct Funct Genom 8(2–3):85–97. https://doi.org/10.1007/s10969-007-9024-5
35. Surjit M, Lal SK (2008) The SARS-CoV nucleocapsid protein: a protein with multifarious activities. Infect Genet Evol 8(4):397–405. https://doi.org/10.1016/j.meegid.2007.07.004
36. Shin GC, Chung YS, Kim IS et al (2007) Antigenic characterization of severe acute respiratory syndrome-coronavirus nucleocapsid protein expressed in insect cells: the effect of phosphorylation on immunoreactivity and specificity. Virus Res 127(1):71–80. https://doi.org/10.1016/j.virusres.2007.03.019
37. Liu J, Sun Y, Qi J et al (2010) The membrane protein of severe acute respiratory syndrome coronavirus acts as a dominant immunogen revealed by a clustering region of novel functionally and structurally defined cytotoxic T-lymphocyte epitopes. J Infect Dis 202(8):1171–1180. https://doi.org/10.1086/656315
38. Gimenez LG, Rojas A, Mendoza J et al (2009) Development of an enzyme-linked immunosorbent assay-based test with a cocktail of nucleocapsid and spike proteins for detection of severe acute respiratory syndrome-associated coronavirus-specific antibody. Clin Vaccine Immunol 6(2):241–245. https://doi.org/10.1128/CVI.00252-08
39. Haynes LM, Miao C, Harcourt JL et al (2007) Recombinant protein-based assays for detection of antibodies to severe acute respiratory syndrome coronavirus spike and nucleocapsid proteins. Clin Vaccine Immunol 14(3):331–333. https://doi.org/10.1128/CVI.00351-06
40. Woo PCY, Lau SKP, Wong BHL et al (2004) Longitudinal profile of immunoglobulin G (IgG), IgM, and IgA antibodies against the severe acute respiratory syndrome (SARS) coronavirus nucleocapsid protein in patients with pneumonia due to the SARS coronavirus. Clin Diagn Lab Immunol 11(4):665–668. https://doi.org/10.1128/CDLI.11.4.665-668.2004
41. Carattoli A, Di Bonito P, Grasso F et al (2005) Recombinant protein-based ELISA and immuno-cytochemical assay for the diagnosis of SARS. J Med Virol 76(2):137–142. https://doi.org/10.1002/jmv.20338
42. Li L, Petrovsky N (2017) Molecular adjuvants for DNA vaccines. Curr Issues Mol Biol 22:17–40
43. Liu S, Wang S, Lu S (2016) DNA immunization as a technology platform for monoclonal antibody induction. Emerg Microbes Infect 5:e33. https://doi.org/10.1038/emi.2016.27
44. Suschak JJ, Williams JA, Schmaljohn CS (2017) Advancements in DNA vaccine vectors, non-mechanical delivery methods, and molecular adjuvants to increase immunogenicity. Hum Vaccine Immunother 12:1–12. https://doi.org/10.1080/21645515.2017.1330236
45. Franconi R, Spanò L, Venuti A, Massa S (2009) Vaccines based on genetic chimera of viral and/or tumoral antigens and plant proteins. European patent EP2456785
46. Massa S, Paolini F, Spanò L et al (2011) Mutants of plant genes for the development of cancer vaccines. Hum Vaccines 7:147–155
47. Franconi R, Massa S, Venuti A (2016) Plant protein signal sequence as adjuvant in DNA vaccines. Patent pending, Submission N. 102016000131935 (29/12/2016)
48. Massa S, Paolini F, Curzio G et al (2017) A plant protein signal sequence improved humoral immune response to HPV prophylactic and therapeutic DNA vaccines. Hum Vaccine Immunother 13(2):271–282. https://doi.org/10.1080/21645515.2017.1264766
49. Broderick KE, Humeau LM (2017) Enhanced delivery of DNA or RNA vaccines by electroporation. Methods Mol Biol 1499:193–200

50. Trimble CL, Morrow MP, Kraynyak KA et al (2015) Safety, efficacy, and immunogenicity of VGX-3100, a therapeutic synthetic DNA vaccine targeting human papillomavirus 16 and 18 E6 and E7 proteins for cervical intraepithelial neoplasia 2/3: a randomised, double-blind, placebo-controlled phase 2b trial. Lancet 386:2078–2088. https://doi.org/10.1016/S0140-6736(15)00239-1
51. Graham RL, Donaldson EF, Baric RS (2013) A decade after SARS: strategies for controlling emerging coronaviruses. Nat Rev Microbiol 11(12):836–848. https://doi.org/10.1038/nrmicro3143

Chapter 11
Panic Disorder During a Bioterroristic Attack

Milan Latas

Abstract There are two ways of viewing panic in the context of bioterrorism: (1) Panic as a social-psychological problem which is present in a large group of people who interact, (2) Panic as an individual psychiatric or psychological problem which is present: (a) in persons with previous panic disorder, or (b) in persons without previous panic disorder.

Panic As a Social-Psychological Problem It is well known fact that humans are vulnerable to high intensity emotions like panic. Panic is often considered infectious in the sense that one person's panic may easily spread to other people around and after that it could be expected that the entire group acts panicky/irrationally very soon.

Panic as an Individual Psychiatric Problem Panic as an individual psychiatric problem is the main focus of the medicine and psychiatry and it is the main focus of this paper. There are several aspects of panic in this context – panic as an individual psychiatric problem. These are (1) panic attack; (2) panic disorder and (3) phobia as the consequence of panic.

Panic Disorder in the Context of Bioterrorism One of the possible targets of this objective is to produce panic in vulnerable subgroups of people with expectation that they will spread it through population. Persons with anxiety disorders, especially persons with panic disorder could be this possible target. In the context of bioterrorism, the question is: what happens after the terrorist attacks with patients with previous panic disorder? The logical hypothesis should be that patients with panic disorder would be more anxious and that they will act with more panic in situation of bioterroristic attack.

But, the results of two research studies, contrary to some expectations, indicate that the presence of real danger does not seem to be associated with higher intensity and frequency of the panic attacks and deterioration in functioning of patients with panic disorder. This means that the presence of real danger does not seem to be

M. Latas (✉)
Belgrade University School of Medicine, Clinic of Psychiatry of Clinical Center of Serbia, Belgrade, Serbia
e-mail: Latasm@eunet.rs

V. Radosavljevic et al. (eds.), *Defence Against Bioterrorism*, NATO Science for Peace and Security Series A: Chemistry and Biology,
https://doi.org/10.1007/978-94-024-1263-5_11

associated with deterioration in a functioning of many patients with panic disorder. Mental disorders (depression, suicide, panic attacks) often develop after the war – terrorist act and we could expect the onset of mental problems, not during but after the possible bioterrorist attack.

Panic is a sudden sensation of fear and overwhelming anxiety which is so strong as to dominate people's behavior and prevent rational and logical thinking of a person.

11.1 Etymology

The word panic derives from antiquity and is a tribute to the ancient god Pan. God Pan was the god of woods and pastures and the ancient Greeks believed that he often wandered peacefully through the woods, playing a pipe but when accidentally awakened from his noontime nap he could give a great shout that would cause flocks to stampede. Therefore, the word panic derives from the name of the Greek god Pan who was in the habit of frightening humans and animals.

11.2 Panic in Context of Bioterrorism

There are two ways of viewing panic in the context of bioterrorism:

1. Panic as a social-psychological problem which is present in a large group of people who interact, and
2. Panic as an individual psychiatric or psychological problem which is present:
    (a) in persons with previous panic disorder, or
    (b) in persons without previous panic disorder.

11.2.1 Panic as a Social-Psychological Problem

It is well known fact that humans are vulnerable to high intensity emotions like panic. Panic is often considered infectious in the sense that one person's panic may easily spread to other people around and after that it could be expected that the entire group of people acts panicky/irrationally very soon. And that is the fact in the large groups of people in crowd place. Panic, as the huge intensity of anxiety and irrational thinking, could lead to irrational behavior, sometimes with unnecessary casualties.

The examples of mass panic are stampedes at the football stadiums, such as stampede at the Heysel stadium in Belgium in 1985 with more than 600 casualties, including 39 deaths and stampede at the Hillsborough stadium in Sheffield, England, in 1989 when 96 people were killed. And there are so many such examples through the world. With this in mind, the possible bioterroristic planers could expect that the people would act panicky in the possible bioterroristic operation and that could be one of the reasons for such action. With expectation of mass panic planers of bioterroristic act could get few objectives in one action – many casualties, lot of attention, onset of mass fear and anxiety after the act etc.

With this in mind it would be important to prepare individuals and society for the possibility of the bioterroristic act and the possible onset of the mass panic in the large groups of people [16]. However, psychiatry and psychiatrists do not deal with mass panic in the large groups of people and this important issue is part of social and psycho-social studies. Psychiatry and psychiatrists deals with panic as an individual problem and this will be the main focus of this chapter.

11.2.2 Panic as an Individual Psychiatric Problem

Panic as an individual psychiatric problem is in focus of the medicine and psychiatry. There are several aspects of panic in the context of panic as a psychiatric problem in individuals. These are: (1) panic attack; (2) panic disorder and (3) phobia as the consequence of panic.

11.2.2.1 Panic Attack

Panic attack is a sudden period of intense fear which may include [1]:

(a) somatic symptoms, such as palpitations, sweating, shaking, shortness of breath, or
(b) feeling (cognition) that something bad is going to happen – going crazy or dying from a heart attack.

The maximum degree of symptoms typically occurs within minutes and symptoms, typically, last for about 30 min. The duration of the attacks can vary from seconds to hours depending on various factors.

It should be noted that panic attacks are not unique to panic disorder. Panic attacks may occur in any of the anxiety disorders, on exposure to feared events – such as in panic disorder or in phobias. Also, panic attacks may occur in other psychiatric disorders, such as exposure to a social situation (like in social phobia) or exposure to a trauma cue (such as in post traumatic stress disorder), after the illegal drug use (like cannabis) and during some medical problems (prolapse of mitral valve). Moreover, panic episodes may be reported by individuals not meeting criteria for any specific mental disorder and they are, mostly, occurring during the stressful situations (like possible bioterroristic act).

11.2.2.2 Panic Disorder

Panic disorder is a specific psychiatric syndrome which is characterized by [1]:

1. recurrent, unexpected, persistent panic attacks or
2. history of one panic attack accompanied with feeling of severe anxiety about having another attack or feeling of severe anxiety about consequences of the attack (such as losing control, having a heart attack, "going crazy") or significant maladaptive behavioral changes as the consequence of the panic attacks – such as avoidance of various situations, going to emergency services etc. [17]

11.2.2.3 Agoraphobia

The term agoraphobia refers to a patient's fear, discomfort, or avoidance of situations in which escape may be difficult or help may not be available in the event of a panic attack [1]. The typical agoraphobic situations may include: traveling on public transportation, going over the bridges, being in open spaces or crowded places, standing in lines etc., and all of this cause various disabilities in functioning of the patients [7, 12].

11.3 How to Recognize Panic?

Like other psychiatric conditions, panic disorder is often unrecognized and untreated. It is usually misdiagnosed as a medical condition (such as thyroid disease, cardiac arrhythmias, obstructive pulmonary disease…).

Therefore, if we want to adequately recognize panic and panic disorder and distinguish them from the other mental and somatic disorders, we should focus on the main features at the disorder [20]. They are: (a) Sudden onset, (b) Number of somatic sensations and irrational cognitions… (like palpitations, tachycardia, shortness of breath, fear of dying or losing control etc.); (c) Huge intensity of fear; (d) Irrational behavior (like immediately escaping from the situation or going to emergency health service).

Also, panic attacks and panic disorder should be differentiated from various somatic illnesses [6]. This raises the question: How to distinguish panic attack and somatic illnesses with similar clinical manifestations? The proper answer to this question should be that sometimes it is very difficult to differentiate those conditions, and it is sometimes very easy to make a mistake in diagnostic procedure [10].

In one analysis related to this question authors have done the research if panic attacks could simulate presence of somatic illnesses [8, 9]. They have analyzed various symptoms of panic attacks and the results of the analyses of the symptoms of panic attacks point to their intercorrelation. This means that, for example, symptoms of cardio-vascular system always correlate with each other. Therefore, this specific association of the symptoms if they are examined individually in the patients with panic attacks without observing the whole problem and other symptoms presented in the patient could lead to false diagnosis of some somatic illness instead of a panic attack. So, it is necessary to analyze all symptoms of the disorder adequately and make proper differential diagnosis of patients with panic disorder.

11.4 Treatment of Panic and Panic Disorder

After the adequate distinguishing procedure, the next question is: What to do or how to cure? The proper answer is that there are different approaches to acute panic attack and to chronic panic disorder.

In acute panic attack the main interventions should be focused on: (1) reassurance of the patient in a calm manner because most panic attacks spontaneously resolve within 30 min; (2) establishing the normal breathing pattern by instructing the patient to breath more abdominally using the diaphragm. If the symptoms of the panic attacks are severe and distressing the use of benzodiazepines could help by bringing an immediate relief of anxiety which could help to reassure the patient, to provide confidence that treatment is possible, and to reduce subsequent emergency presentations of the patient [15]. In case that patient has not had the attacks in his personal history and if it is the first presentation of the attack, the physician should exclude medical causes of the attack. This procedure may require admission to hospital for specific examinations.

In the other cases – in chronic panic disorder, which means that the panic attacks are recurrent and frequent, the physician should consider differential diagnosis for panic disorder and address of the underlying psychiatric disorder may require psychiatric referral. In addition, the goals of the treatment of panic disorder are to significantly reduce or eliminate mean features of the attacks: (1) panic attacks, (2) phobic avoidance behavior, and (3) anticipatory anxiety – fear about next panic attack.

First line therapy for the patients of panic disorder should be pharmacotherapy or/and psychotherapy [2, 7, 12, 13]. The pharmacotherapy should be based on the use of antidepressants, mainly on the use of specific group of selective serotonin reuptake inhibitors – SSRIs (paroxetine, fluoxetine, sertraline, citalopram and escitalopram) or benzodiazepines, especially alprazolam and clonazepam [11, 14]. The psychotherapy could be provided as the only treatment for the patients but, it could be, also, provided in combinations with drugs [11, 14]. The main psychotherapy modalities that could be used efficiently in patients with panic disorder are: cognitive-behavioral therapy (CBT), psychodynamic, and interpersonal psychotherapy. The main therapy method for the phobias related to panic disorder is cognitive-behavioral therapy (CBT) with exposure to phobic stimuli as the most important therapeutic procedure [21].

11.5 Panic Disorder in the Context of Bioterrorism

One of the main objectives of the planers of terroristic attacks is to get people to feel fear and anxiety, to worry about their lives and the lives of their family and friends [5]. Moreover, the objective of terroristic attack is that the people should get panic as the attack is unpredictable and because it could have large consequences concerning life and health [4]. One of the possible targets of this objective is to produce panic in vulnerable subgroups of people with expectation that they will spread it through population. Persons with anxiety disorders, especially persons with panic disorder could be this possible target.

Regarding bioterrorism attack and its consequences to the persons with panic disorder it is rational to think that persons who are more prone to anxiety (like persons with panic disorder) will act with more concerns and fears that other people in population. It should be expected that someone with panic disorder should behave with more panic in this particular situation. This means that in dangerous situation, like terroristic attack, people with panic disorder will have more symptoms of anxiety and that their behavior should be more irrational. On the other hand, we could assume that the successful managing of panic disorder lead to greater resistance in vulnerable population. This means that persons with panic disorder which is in remission and in stable phase could not be good target for the bioterrorism attack in a sense of producing and spreading fear and panic.

However, the results of the studies pointed out that patients with panic disorder usually shift their focus of attention from external to internal stimulus [3, 8, 9]. This means that patients with panic disorder are mainly interested in somatic sensations – the symptoms of the panic attack and that they are much less interested in other stimuli. As an example, person with panic disorder focus his attention to the symptoms of possible panic attack, such as tachycardia or dizziness and defocus his attention from other internal and, especially, from external stimuli. This indicate that other stimulus, like life events, are not in central attention of the person because of the fear of the possible onset panic attack and its life dangerous consequences.

This raises the question: what happens when external stimuli are strong enough to attract attention of most people? For example, wars and terroristic attacks are situations in which external stimuli can hardly be ignored. With this in mind, the question could be: are patients with panic disorder able to pay adequate attention to real danger and to interpret it correctly? In the context of bioterrorism, the question should be: what happens during and after the bioterrorist attacks with patients with panic disorder?

The logical hypothesis should be that patients with panic disorder would be more anxious and that they will act with more panic in situation of bioterroristic attack. But there is contrary hypothesis based on anecdotal reports and study results which suggest that patients with panic disorder are able to pay adequate attention to real danger and to interpret it correctly, unlike their tendency to misinterpret internal – somatic stimuli “catastrophically” ([19]).

If we want to get the proper answers we should focus on the results of studies that examined issues about relationship between panic disorder and terroristic attack – are there more symptoms of anxiety and panic in persons with previous panic disorder and are these persons prone to spread panic through population in a situation of bioterroristic attack. Unfortunately, there are no such studies, i.e. studies that examined behavioral patterns and cognitions in patients with panic disorder in the context of bioterroristic attacks. However, there are some studies related to war period, periods during and after the missile attacks to civilian targets, which assessed various aspects of the emotions, cognitions and behavior of persons with panic disorder who were present in such situations. With the results of these studies we could assume and may be predict possible reactions of panic-disordered persons in circumstances of bioterroristic act.

11.5.1 Panic Disorder and War Period

Authors from Israel conducted the research with the aim to systematically assess the behavioral effects of a real, life-threatening event on panic-disorder patients during the missile attacks on civil targets [18]. The authors assessed 65 patients with panic disorder. Those patients completed interviews during the Persian Gulf War and evaluation included frequency of panic attacks, level of anxiety and function levels of the patients. The evaluation of the patients was carried out during air raid alarms (which was perceived as dangered period) and between air raid alarms (which was perceived as a period without dangeros).

Contrary to expectations, the results of this study indicate that patients with panic disorder, despite high levels of anxiety, did not demonstrate an increased frequency of panic attacks during the war period specifically during the missile attacks. In addition, the authors discovered that the majority of patients reported good to high levels of functioning. This was obvious in both everyday functioning and functioning during the alarm-related period. The authors of the study concluded that vulnerability of patients with panic disorder to a "panic-stricken" response does not increase during real-life stressors and that the lack of increased frequency of panic attacks observed under these circumstances provides additional support for the opinion that panic and fear are two distinct entities.

In the other study the objective was to assess various aspects of panic disorder in the context of real danger situation [19]. The question was: how do patients with panic disorder respond to unsafe situation and how do they cope in such situations? The authors investigated the panic disordered patients who were exposed to air strikes against Yugoslavia in 1999 where the bombing campaign was mainly directed at military targets but were the civilian targets were also hit and where were civilian casualties. They sought to examine frequency and intensity of: (1) panic attacks, (2) anticipatory anxiety, (3) agoraphobic avoidance, (4) panic-related health concerns and (5) level of disability of the patients all during the war period in comparison to pre war.

The results of this study show that the patients with panic disorder have had greater intensity and frequency of anticipatory anxiety but decreased overall severity of panic disorder, almost the same intensity and frequency of panic attacks and almost the same intensity of agoraphobic avoidance, fewer health concerns and decrease in the level of disability, all comparing to period before the air strikes.

The results of those two studies, contrary to some expectations, indicate that the presence of real danger does not seem to be associated with higher intensity and frequency of the panic attacks and deterioration in functioning of patients with panic disorder. The results of those studies pointed out that the patients with panic disorder can be reassured that they are not likely to cope worse under conditions of danger. The results of those studies also imply that panic disordered patients will not act panicky and that they will not spread panic to the others.

11.5.2 Speculations About Panic Disorder and Bioterrorism Attack

With aforementioned in mind we could hypothesize that the patients with panic disorder will not fear and behave with more panic and fear in the hypothetical situation of bioterrorism event. This could imply that those patients will not be vector of the dissemination of fear and panic to the others.

In addition, we could speculate on the implications of the differences between the types of war-related events on patients with panic disorder. Some of them are characterized by sudden violence, usually over very quickly – like a missile attack and some of them are characterized by the slowly unfolding character – like bioterrorism event. The results of the previous mentioned studies imply that patients with panic disorder will not get more panic in the situation of sudden violence. On the basis on this information we could assume that patients with panic disorder in the situation of bioterrorism incident will not act panicky regardless gradually unfolding of the event. This assume is based on the fact that in real danger situation patients with panic disorder adequately shift their attention and behave rational despite the potential danger event.

On the other hand, mental disorders (such as depression, suicide, panic attacks) often develop after the war – in this case the terroristic act [22]. As a result, we could expect the onset of mental problems, not during but after the possible bioterrorist attack, especially in vulnerable population – people with previous mental disorders, without social network, with previous psychological or social problems etc. This could be the same in patients with panic disorder. After the bioterrorist event we could expect the worsening of mental health status in those patients – with more panic attack, more anticipatory anxiety and more phobic avoidance. Also, after the possible terroristic event we could expect more new onsets of panic disorder in the population affected by the event.

11.5.3 What to Do?

All of this raises the question: How to increase the psychological resilience in the context of bioterrorism attack? Preventive approaches should focus on:

- helping the mentioned vulnerable groups, such as peple with previous mental disorders or disadvantaged groups that lack social and economic support,
- strengthening community solidarity and
- strengthening confidence in the local leadership.

All of this is important because the "sense of belonging" is one of the most powerful factor of resilience, so the leaderships should be educated and trained what to do in case of possible terroristic acts to give adequate help to people who need it. However, in the context of bioterrorism event and its implications on patients with panic disorder and people who are prone to develop panic disorder the first activity is the most important. This activity should include screening of the vulnerable individuals, support groups for vulnerable individuals with trained professionals, and referring the individuals with obvious mental problems to mental health professionals. Those activities should be organized not during the bioterrorism event but, mandatory, after the bioterrorism event when is the most risk period for developing mental disorders in exposed and vulnerable individuals.

References

1. American Psychiatric Association (2013) Diagnostic and Statistical Manual of Mental Disorders (DSM-5). American Psychiatric Association, Washington, DC
2. Bandelow B, Lichte T, Rudolf S, Wiltink J, Beutel ME (2015 Aug) The German guidelines for the treatment of anxiety disorders. Eur Arch Psychiatry Clin Neurosci 265(5):363–373
3. Clark MD (1986) A cognitive approach to panic disorder. Behav Res Ther 24:461–470
4. Goel AK (2015 Jan 16) Anthrax: a disease of biowarfare and public health importance. World J Clin Cases 3(1):20–33. https://doi.org/10.12998/wjcc.v3.i1.20
5. Jansen HJ, Breeveld FJ, Stijnis C, Grobusch MP (2014 Jun) Biological warfare, bioterrorism, and biocrime. Clin Microbiol Infect 20(6):488–496. https://doi.org/10.1111/1469-0691.12699
6. Latas M, Adnadjevic S (2007) Relation of panic disorder and somatic illnesses. Eur Psychiatry. 22(Suppl. 1):S279
7. Latas M, Milovanović S, Backović D (2004a) Antidepressants as a monotherapy option in patients with panic disorder. Pharmaca Iugoslavica 42(1–2):20–25
8. Latas M, Obradović D, Pantić M (2009a) Fear of somatic sensations and symptoms in persons with panic disorder or somatic illnesses. Srp Arh Celok Lek 137(11–12):659–663
9. Latas M, Soldatović I, Stamenović M, Starcević V (2009b) Panic attacks simulate presence of somatic illnesses. Med Pregl 62(3–4):165–170
10. Latas M, Starčević V (1998) Comorbidity in psychiatric disorders – theoretical considerations. Psihijatrija Danas 30(1):127–136
11. Latas M, Starčević V, Trajković G, Vučinić D (2002) Combination of benzodiazepines and SSRIs in treatment of panic disorder. Eur Psychiatry 17(Suppl. 1):162s
12. Latas M, Starcevic V, Vucinic D (2004b) Predictors of work disabilities in patients with panic disorder with agoraphobia. Eur Psychiatry 19(5):280–284

13. Latas M, Stojanović D, Batinić B, Mamula MD (2003) Current treatment of anxiety and phobic disorders. Engrami 25(3):15–24
14. Latas M, Vučinić D (2002) The efficacy of CBT treatment in patients with panic disorder. Engrami 24(2–3):55–61
15. Latas M (2006) Comparison of high-potency benzodiazepines vs. combination of high potency-benzodiazepines and fluoxetine in treatment of panic disorder with agoraphobia. Int J Neuropsychopharmacol 9(Suppl. 1):S195
16. Radosavljevic V, Radunovic D, Belojevic G (2009 Sep) Epidemics of panic during a bioterrorist attack–a mathematical model. Med Hypotheses 73(3):342–346. https://doi.org/10.1016/j.mehy.2008.12.055
17. Roy-Byrne PP, Cowley DS (1995) Course and outcome in panic disorder: a review of recent follow-up studies. Anxiety 1:151–160
18. Sasson Y (1999) Response to missile attacks on civilian targets in patients with panic disorder. J Clin Psychiatry 60(6):385–388
19. Starčevic V, Latas M, Kolar D, Bogojević G, Kelin K (2002) Panic disorder patients at the time of air strikes. Depress Anxiety 16:152–156
20. Stein MB, Roy-Byrne PP, McQuaid JR, Laffaye C, Russo J, McCahill ME et al (1999) Development of a brief diagnostic screen for panic disorder in primary care. Psychosom Med 61:359–364
21. Tsao JCI, Mystkowski J, Zucker B, Craske MG (2002) Effects of cognitive-behavioral therapy for panic disorder on comorbid conditions: replication and extension. Behav Ther 33:493–509
22. Wessely S, Hyams KC, Bartholomew R (2001 Oct 20) Psychological implications of chemical and biological weapons. BMJ 323(7318):878–879

Chapter 12
Safety and Security Regulations Against Biological Threats

Anna Bielecka-Oder

Abstract Biological threat agents include bacteria, viruses, fungi, parasites and their associated toxins. They have the ability to harmfully affect human health ranging from an allergic reactions to serious illnesses, even death. Water, soil, air, plants or animals can be a suitable habitat for their live and proliferation.

Because biological agents may reproduce rapidly and initially unnoticed, need minimal resources to survive and can infect at very small doses they can be used as biological warfare agent or bioweapon. Genetic modification may enhance their hazardous and lethal properties, or develop resistance to conventional treatments.

In effect, to protect people from dangerous biological agents as well as protect biological agents from intentional malicious acts both, biological safety and biological security measures should be implemented and respected. Because of wide scale of risks caused by biological agents, biosafety and biosecurity issues should be interpreted on many fields taking as priority protection of human beings and their surrounding environment.

The reader will familiarize with different point of views on biosafety and biosecurity issues in relation to occupational health and safety, public health and disease surveillance, biodiversity protection, genetic modification of microorganisms, transportation of dangerous goods, storage control of biological agents, dual-use technology, education and awareness raising, weapon of mass destruction threats and bioterrorism acts.

The main international agreements, the European Union regulations and principles supporting implementation of national legislation concerning biosafety and biosecurity areas are presented. The role of legally and not-legally binding instruments is highlighted, as well.

A. Bielecka-Oder (✉)
The Analysis Team, The Epidemiological Response Center of the Polish Armed Forces,
7 Leskiego Street, 01-485, Warsaw, Poland
e-mail: anna.bielecka@poczta.fm

V. Radosavljevic et al. (eds.), *Defence Against Bioterrorism*, NATO Science for Peace and Security Series A: Chemistry and Biology,
https://doi.org/10.1007/978-94-024-1263-5_12

Keywords Biosafety · Biosecurity · Regulations · Guidelines

12.1 Biological Threats

Biological threats contain biological material planned to be deployed to affect human community and/or environment providing human and animal diseases or else damage of plants [1]. Biological agents are sorted into three basic groups that would likely to be used as biological weapon – bacteria, viruses and toxins. Examples of high risk bacterial threats include among others anthax *(Bacillus anthracis),* plague *(Yersinia pestis),* epsilon (alpha) toxin of *Clostridium perfringens* and tularemia *Francisella tularensis).* Examples of high risk viral threats includes influenza, coronaviruses (e.g. MERS and SARS), smallpox and Ebola virus. Examples of high risk toxins include botulinum neurotoxin and ricin. Moreover, biological agents comprise new and re-emerging pathogens as well as intentionally genetically modified infectious agents [2]. A biological threats are typically associated with the deliberate release of a biological agent in an act of terrorism. However, such threat can be a result from an accident in laboratory, as well.

The Valuable Biological Material (VBM), which includes biological agents requires administrative oversight, control, accountability and specific protective and monitoring measures in laboratories to maintain their economic and historical value and/or the population from their potential to cause harm. The VBM may include pathogens and toxins as well as non-pathogenic organisms, vaccine strains, food, genetically modified organisms (GMOs), cell components, genetic elements and extraterrestrial samples [3].

The diversity of biological agents and pathogens makes their management a significant challenge. In consequence both, biological threats and high consequence pathogens always deserve special attention.

12.2 Delivery Methods of Biological Agents

Contact transmission is the most common form of bacteria and viruses dissemination. There are two types of contact transmission: direct and indirect.

Direct contact transmission occurs when there is physical contact between an infected person and a susceptible person. Types of direct contact include both, (1) person-to-person interaction when an infected person comes into direct connection via touches or exchanges body fluids with someone else, and (2) droplet spread during coughing, sneezing and speaking, when droplets fall to the ground within a few feet and infect in close proximity.

Infectious diseases can also be spread indirectly, when there is no direct human-to-human contact. For example thru (1) airborne transmission, (2) contaminated objects, (3) food and drinking water, (4) insect bites or animals and (5) environmental reservoirs.

Some biological agents travel long distances and remain suspended in air for an extended period of time. Inhaling biological agents dispersed into the air (bioaerosols) may cause disease in people or animals indirectly, e.g. cold, influenza.

Furthermore, some microorganisms can live on objects for a short time. Mode of transmission via contaminated objects occurs when you touch an object such as a doorknob or bedding soon after an infected person, done it before you. It is high probability that you might be exposed to an infection, e.g. smallpox.

Infectious diseases can be also transmitted by food and water. *Salmonella* sp. is often transmitted through improperly handled produce or undercooked poultry meat. Improperly canned foods can create an environment ripe for *Clostridium botulinum*, which can lead to botulism.

Additionally, indirect spread of biological agents can happen in way from an animal reservoirs to vectors. Zoonosis occurs when disease is transferred from an animal to human. Zoonotic diseases include inter alia anthrax (from sheep), rabies (from rodents and other mammals), West Nile virus (from birds) and plague (from rodents). Besides, some zoonotic infectious agents may be transmitted by insects bites, especially those that suck a blood. These include mosquitos, fleas and ticks. The insects become infected when they feed on infected hosts previous, such as birds, animals and humans. The West Nile virus, Zika virus, tick-borne meningitis and Lyme disease are all spread by this way.

Moreover, soil, water and vegetation containing infectious microbes can also be transferred to human beings. Legionnaire disease is an example of a disease that can be spread by water.

In reference to biological attack and presented delivery methods our attention should paid unnatural pattern of disease transmission, reserve zoonosis spread (human to animals) or direct evidence of agents release. Based on that an appropriate prevention methods should be evaluated.

12.3 Definition of Biosafety and Biosecurity

Biological safety or biosafety includes measures aimed at protecting people and the environment from the unintentional impact of biological agents. It is rather related with prevention of incidental exposure, unintentional release, or accidental loss. Biosafety is mostly related to human, animal and plant health protection [4, 5]. Biosafety mostly relay on both, (1) occupational health and safety as well as (2) genetically modified organisms (GMOs) and biodiversity protection. Nevertheless, it also takes into account medical health care, epidemiology, agriculture, rDNA experiments, biosecurity and the Biological Weapon Convention (BWC) issues.

Biological security or biosecurity is much younger concept than biosafety [6] and comprises measures that minimize the possibility of biological agents being deliberately used to cause harm. It is rather related with prevention of theft and unauthorized access, intentional misuse or diversion, dual-use issues, bioterrorism, intentional illegal release or criminal event. Biosecurity is more complex because has different associations in many contexts. From public health point of view biosecurity is much more related to security and oversight of pathogenic microorganisms and toxins in both, microbiology facilities and during transfer/transport. Biosecurity mostly is focusing on "protection, control and accountability for valuable biological materials within laboratories, in order to prevent their unauthorized access, loss, theft, misuse, diversion or intentional release [7]." In veterinary and agricultural sectors the biosecurity definition has come to denote protecting biological resources from foreign (not naturally occurring in this region, including genetically modified) or invasive species [8]. Moreover, biosecurity close corresponding to legal, regulatory and administrative measures export-import control of dangerous goods, facility-based biosecurity, security of clearance for personnel, intellectual rights protection, biosecurity of dual-use goods, crime investigation of biological threats, international agreements relevant for the non-proliferation of biological materials, equipment and technology (e.g. BWC, EU Green Paper, UN Resolution 1540), terrorism, state security issues, and of course biosafety subjects.

Both, biosafety and biosecurity are multi-source related and derived. Beside, defined in different ways by various authorities are always strictly related to biological agents [9].

During the Meeting of States Parties to the Biological and Toxin Weapons Convention (BWC) in 2003 the informal terms for biosafety and biosecurity were suggested. It was said that "Biosafety protects people from germs, and biosecurity protects germs from people [10]." Despite these definitions are informal, they present the merits of biological safety and biological security in up to the point and in easy way.

12.4 Legal Framework on Biosafety and Biosecurity

Legal framework on biosafety and biosecurity include both, legally and non-legally binding instruments.

Legally binding instruments carry the force of law and require signatories to comply with the agreements as adopted. This may include ratification, accession and/or transposition of agreements into national frameworks through implementation process. Legally binding instruments include international agreements and conventions, the European Union regulations and, different legal and constitutional arrangements in homeland legislations.

Compared with binding agreements, non-legally binging instruments do not create binding obligations and are not legal instruments enforceable by the national institutions. Consequently, there is no formal requirement to adopt them into national legislation. The advantage of non-binding agreements is being faster and simpler to adopt than binding agreements and providing more flexible means for update and adjustment. Non-legally binging instruments represent standards, guidelines, manuals, codes of conduct, good practices, recommendations and/or declarations, which are dealing with professional issue.

12.4.1 International Agreements and Conventions

International instruments regulating biosafety and biosecurity matters include treaties, conventions and agreements apply to most countries, but not always all. A number of existing agreements have been launched and implemented by UN agencies, although not all its members are signatories or parties of them.

The Convention on the prohibition of the development, production, and stockpiling of bacteriological (biological) and toxin weapons and on their destruction or biological weapons convention (BWC) is the first and the most important treaty banning the use of biological weapon and promoting biosafety and biosecurity on a global scale.

The Cartagena Protocol on Biosafety to the Convention on Biological Diversity (Cartagena Protocol) is addressing some environmental and health impacts of modern biotechnology. From biosafety and biosecurity point of view, the Protocol regulates an international transport and release of genetic modified organisms (GMOs) to protect natural biological diversity. The treaty was adopted in January 24–28, 2000.

The next one, the UN Security Council Resolution 1540 (Resolution 1540) passed in 2004, calls to all UN members to give high priority of joining and implementing the non-proliferation treaty regimes including the BWC.

Also, WHO International Health Regulations (IHRs) revised in 2005 represents a binding international legal agreement involving 196 countries across the globe. It aims to help State Parties in capability-building to prepare and respond to natural, accidental and intentional spread of diseases as well as to improve the BWC's Confidence Building Measures (CBMs) information exchange practice. Analogous to the Resolution 1540, the IHRs requires to consider and adopt both, biosafety and biosecurity regulations and practices.

12.4.1.1 The Biological Weapon Convention (BWC)

The first Article of the BWC presents the aim and scope of the whole treaty, namely, each State Party should take all necessary safety and security measures to protect the population and the environment as well as take into account scientific and technological achievements in the field of microbiology, genetic engineering, biotechnology, molecular biology and synthetic biology [11].

The biosafety concept is openly cited in the Article II, which requires States Parties to "destroy, or to divert to peaceful purposes" any biological weapons they have and specifies that in implementing this requirement "all necessary safety precautions shall be observed to protect populations and the environment [12]."

Biosecurity concept under the Convention is refered in Articles III and IV. The Article III calls State Parties to implement legal regulations in the field of transfer biological agents and measures indicated in the Article I of the Convention for peaceful purposes, including their security and protection during transportation [13]. The spirit of the biosafety and biosecurity concepts are evidently presented under the scope of the Article IV of the BWC, which stipulates Member Countries to take "any necessary measures [14]" to prohibit and prevention the proliferation of weapons of mass destruction (WMDs), encompassing according to the Final Declaration of the Sixth Review Conference legislative, administrative, judicial and other measures, including penal legislation. Through sentence "any necessary measures" should be also understood national implementation of the BWC, ensuring the safety and security of microbial or other biological agents or toxins in laboratories, facilities, and during transportation to prevent unauthorized access to and removal of such agents or toxins as well as education and raising awareness on BWC [15].

International consultation and effective cooperation of States Parties are a welcome tools for promoting biosafety and biosecurity practices, as well (Article V and VI of BWC [16].

The following, a mutual assistance and support of both, domestic agencies of State Parties and international organizations (e.g. WHO, INTERPOL, FAO, OIE, IPPC), in case of bioterrorism act suspicion or in response of any biological threat also will enhance global biological safety and security [17]. The Article VII promotes assistance in both, epidemiological and criminal investigations.

Finally, the possibility of using scientific knowledge for peaceful or malicious purposes reflects a dual-use dilemma and affects both, publication of knowledge and the biological agents themselves. Biosecurity from this point of view may be defined as a "balancing the promise of biotechnology with preventing the biological weapons threat [18]." The Article X of the Convention endorses cooperation of State Parties in transfer of knowledge and technology for peaceful uses of biological sciences. Therefore, it calls for closer institutional and educational cooperation between nations in order to improve protection of human and animal health [19].

Full implementation of the BWC is an important contribution to the fight against bioterrorism. Most, but unfortunately not all countries (Chad, Comoros, Djibouti, Eritrea, Israel, Kiribati, Micronesia (Federated States of), Namibia, Niue, South Sudan and Tuvalu) are signatories of the convention.

12.4.1.2 The Cartagena Protocol on Biosafety

The Article I of the Cartagena Protocol states that the main aim is "contribute to ensuring an adequate level of protection in the field of the safe transfer, handling and use of living modified organisms (LMOs) resulting from modern biotechnology that may have adverse effects on the conservation and sustainable use of biological diversity, bearing in mind risks to human health, and specifically focusing on transboundary movements [20]."

In brief, the Protocol promotes biosafety by establishing rules and procedures for safe trade of LMOs in order to protect natural biological diversity.

12.4.1.3 UN Security Council Resolution 1540

The Resolution 1540 (followed by the Resolution 1673 in April 2006) describes provisions for prevention of proliferation of nuclear, chemical and biological weapons. The main obligations are contained in the first three operative paragraphs including (1) prohibit States to provide "any form of support to non-State actors that attempt to develop, acquire, manufacture, possess, transport, transfer or use (…) biological weapons and their means of delivery", (2) to adopt and enforce (…) laws to prohibit such activities under their national legislation and (3) implement and enforce a comprehensive system of domestic controls on WMD and related materials [21].

In summary, the Resolution 1540 promotes biosecurity concept by establishing domestic laws for secure handling of biological material, physical protection of facilities where "high emergency pathogens" are located as well as border controls of biological material, means of delivery and dual-use items.

12.4.1.4 International Health Regulations

The main role of the IHRs, an international legal instrument drafted by the World Health Organization (WHO) is to enhance worldwide public health security. This role should be pursued through prevention, protection against, control and response to the international spread of disease, strengthen the collective defenses against the multiple and varied public health risks and finally, setting of rules to support the global outbreak alert and response system in order to improve international surveillance and reporting mechanisms for public health events, strengthen their national surveillance and response capacities.

Additionally, the WHO IHRs applies to all spectrum of public health risks, from natural disease outbreaks, new or re-emerging diseases, unintended consequences, then accidents, negligence, vandalisms or sabotages until deliberate use of biological weapon [22].

By adoption of the WHO IHRs, States Parties become more aware of the connection between public health and biological weapons reflecting closer integration the WHO and the BWC on biosafety, biosecurity, disease surveillance and reporting actions.

12.4.2 EU Legal Instruments on Biosafety and Biosecurity

Among regional instruments, the European Union (EU) legislation is one of the most extensive. The EU provides a common regulatory framework for biosafety and biosecurity issues mainly through directives adopted by the Council of the European Union and the European Parliament. All EU directives require implementation by EU member states through their national legislation. Therefore, the national regulations on biosafety or biosecurity may look different in the EU countries, but ought to share a common minimum standard.

The most important EU regulations in biosafety and biosecurity fields are represented by both, the Directive 2000/54/EC on protection of workers from risks related to exposure to biological agents at work [23] and the Directive 2009/41/EC related to contained use of genetically modified microorganisms (GMMs) [24].

The Directive 2000/54/EC provides minimum requirements for occupational health and safety of workers exposed to biological agents beginning from the basic definitions in this field. Following, the Directive classifies the biological agents into four groups of threats based on the level of posted risk. It also provides information on how to prevent exposure to several groups of biological agents and limit the risk. Moreover, responsibilities of employers are described with respect to work involving (or likely to involve) exposure to biological threats. This Directive obligates each State Party to establish national provisions for carrying out surveillance of worker's health care prior to exposure and at regular intervals thereafter. The employer is obliged to ensure effective vaccines free of charge for his personnel. If a worker is found to be suffering from an infection or illness as a result of exposure at workplace, additional diagnostic tests should be offered similarly to other co-workers. Besides particular attention should be paid on suspicious presence of biological agents in human patients and animals, hazards represented by biological agents present in human patients or sick animals, and risks posed by the nature of the work. Appropriate decontamination and disinfection procedures should be implemented at workplace. Contaminated wastes should be handled and disposed without harm. Laboratories handling or use biological agents classified at 2, 3 or 4 group (according to the Article 2 and the Annex VI of the Council Directive 90/679/EEC) need to implement relevant containment measures in order to minimise the risk of infection.

As well as occupational safety systems, environmental protection needs to be considered in biological safety and security. Modern biotechnology is an emerging science with a great potential not only in improving human and animal health, ensuring the maintenance of biodiversity and environmental protection, agriculture, industrial and agricultural production, but also probability of using new scientific knowledge for malicious purposes. The EU has established a legal framework for safe and secure development of GMOs products based on both, (1) human and animal health, environmental protection, as well as (2) clear labelling and traceability of GMOs placed on the market.

The first one aspect, human, animal, and environmental health protection, is realized by the Directive 2009/41/EC on the contained use of genetically modified microorganisms (GMMs) [24], the Directive 2001/18/EC on the deliberate release into the environment of GMOs [25], the Directive (EU) 2015/412/EU on cultivation of EU authorized GMOs on their territory [26] and the Regulation (EC) 1829/2003 on genetically modified food and feed [27].

The Directive 2009/41/EC is the most important regulation on contained use of GMMs. In order to ensure a high level of GMMs protection, the containment levels (1–4 classes) and other protective measures should be used. Moreover, it recommends using adequate safety and security measures in laboratories involving administrative procedures and/or notification requirements associated to the risk of the contained use of GMMs. The Annex IV of this Directive presents tables with minimum requirements and protective measures necessary for each level of containment. Additionally, it provides general terms related to this field and presents activities needed to be considered to develop risk assessment. In order to increase the flexibility for the amendment of the technical annexes, allowing timely adaptation to scientific and technical progress, the Commission may assist by advisory committees in necessity. Discussing further directives in this area goes beyond the scope of this publication.

The second aspect, the labelling and traceability of GMOs products is controlled by EU regulations, namely, the Regulation (EC) 1829/2003 on genetically modified food and feed [27], the Regulation (EC) 1830/2003 concerning the traceability and labelling of genetically modified organisms and the traceability of food and feed products produced from genetically modified organisms [28], to end with the Regulation (EC) 1946/2003 on transboundary movements of GMOs [29].

Additionally, protection of biological diversity is realized by various European agreements, including the Birds Directive [30], the Habitats Directive [31], the EU Biodiversity Strategy [32], the EU Forest Strategy [33] and the Convention on International Trade in Endangered Species of Wild Fauna and Flora (CITES) [34], etc.

Another aspect associated to biosafety and biosecurity measures is fact that infectious diseases may be easily spread and reach new regions. A legal agreements on border screening and control contribute to limiting transmission of infectious diseases in other countries. Many guidelines for transport of dangerous goods including safety and security elements have been translated by the EU into international acts and agreements that Member States were obligated to apply. The European Agreement concerning the International Carriage of Dangerous Goods by Road (ADR) [35] classify infectious substances (Class 6.2) and provide secure measures for its transportation (general and specific requirements related to packed, labelling, marking of containers and vehicles, conditions of carriage, loading, unloading and handling, vehicle crews, equipment, operation and documentation, as well as construction and approval of vehicles). The Convention concerning International Carriage by Rail (RID) [36] provides regulations concerning safe and secure rail transport for dangerous goods. Also, the International Air Transport Association (IATA) consisting of 275 airlines, primarily major carriers, representing 117 countries priority is to ensure safety and secure shipments of infectious substances (certain principles, definitions related to biological agents, classification of infectious substances on A and B category (according to the IATA Dangerous Goods Regulations), list of indicative examples infectious substances in each category, packing instructions for infectious substances category A and B acceptable for air transport with a wide range of options for inner, outer and single packaging, labelling, conditions of carriage, loading, unloading, documentation, etc.), biological products, GMOs and GMMs, medical and clinical wastes, infected animals and patient specimens [37, 38]. Adequate regulations on safe and secure transportation or shipment of dangerous goods or hazardous materials by water on vessel have been also established [39]. Additional directives related transportation of dangerous goods are presented below. Safety and security measures related to transport by road, rail and inland waterway are covered by the EU Directive 2008/68 on inland transport of dangerous goods [40]. Next, an uniform procedures for checks on the transport of dangerous goods by road and by rail are provided in, respectively, the Directive 2008/54/EC [41] and the Directive 2004/110/EC [42]. Safe transport of transportable pressure equipment by road and rail is included in the EU Directive 2010/35 [43].

Furthermore, dual-use nature of biological researches, equipment and processes also increase the likelihood of accidents involving high-risk pathogens. Efforts to strengthen biological security by reducing a threat posed by biological agents should require ingenuity and effective export-import control measures. The Regulation 428/2009 setting up a Community regime for the border control of exports, transfer, brokering and transit of dual-use items [44]. Additionally, the Australia Group (AG) supports associated countries to force export measures for control of dual-use goods. It provides control lists of potential dual-use materials including microorganisms and specific elements of specialist equipment dedicated for their large-scale production in bioreactors. The soft nature of this order make possible to update the catalogs by adding new items in response to actual threats. By actions focused on combating proliferation of chemical and biological weapons the Australia Group strengthen the BWC, as well.

In spite of biosafety, there is no legal framework for biosecurity in EU at the moment. This lack is replaced by international strategies, multilateral initiatives or actions organized and funded by EU [45].

One of the first primordial initiatives focused on the growing international security via Chemical, Biological, Radiological and Nuclear CBRN threats reduction policy was the EU Strategy against Proliferation of Weapons of Mass Destruction (WMD) [46] adopted in 2003. The Strategy identified probable ways of dual-use technology and knowledge misuse, which were considered to be "increasing as a result of rapid developments in life sciences". In reference to biological threats, the Strategy relied on all hazards of biological origin as well as highlighted the necessity to improve "the security of proliferation-sensitive materials, equipment and expertise in the EU against unauthorized access and risks of diversion [47]." This Strategy was updated in 2008. The same year, the European Council adopted a Joint Action 2008/307/CFSP [48], which supports the WHO's activities in the area of laboratory biosafety and biosecurity together with the above Strategy framework.

A one year before, in 2007, in order to reduce biological threats via one more way – enhancing preparedness and response activities, the European Commission had prepared the "Green Paper on biopreparedness [49]." This document arises from the consultation process on prevention, protection, prosecution of criminals/terrorists, surveillance, response and recovery aspects all relevant stakeholders represented by Member States and EU authorities. Summarizing, the discussions were focused on how to improve security and prevent of deliberate criminal acts, accidents as well as response to naturally-occurring outbreaks.

Subsequently, the above Green Paper and consultations made around it led to elaboration of the European Union CBRN Action Plan [50]. The EU CBRN Action Plan was intended to reduce a risk of public safety and security, and simultaneously, strengthen CBRN security in the European Union region. The EU CBRN Action Plan includes all hazard threats – from CBRN incidents of accidental, natural and intentional origin to terrorist acts, contributing the implementation of the EU Counter Terrorism Strategy [51]. In relation to enhancing biosafety and biosecurity, the CBRN Action Plan contains four specific actions – prevention, detection, preparedness and response – dedicated for high risk biological agents and toxins. Most of presented biological actions apply to preventive activities like the list of high risk biological agents and toxins (Goal 1 of the Prevention Chapter), enhancement of high-risk materials and facilities security (Goal 2 of the Prevention Chapter) and control (Goal 3 of the Prevention Chapter), participation of workers in the development of a high security culture at workplace (Goal 4 of the Prevention Chapter), improvement of identification, reporting, (Goal 5 of the Prevention Chapter), transport security (Goal 6 of the Prevention Chapter), information exchange (Goal 7 of the Prevention Chapter), provision of specific training (Goal 4 of the Actions applicable to CBRN prevention, detection and response Chapter) and personnel security (Goal 5 of the Actions applicable to CBRN prevention, detection and response Chapter).

No less important and also on the subject of biosafety and biosecurity on EU level is the Instrument for Stability (IfS) [52], which was established in 2006. One of its goal was contribution in capacity building against CBRN threats via full implementation of the BWC, strengthening biosafety and biosecurity capabilities according to the modern standards and providing trainings. In consequence of this initiative, the EU supports realization of several projects in Russia and Central Asia countries. In one of them, titled "Bio-safety and bio-security improvement at the Ukrainian anti-plague station (UAPS) in Simferopol" was engaged a Polish institution, the Military Institute of Hygiene and Epidemiology, which provided trainings in this fields [53]. From March 2014, the Instrument for Stability (IfS) was succeeded by the Instrument contributing to Stability and Peace (IcSP) [54].

One of the current conducting EU initiatives is the EU CBRN Risk Mitigation – Centres of Excellence (CoE) [55]. The rationale of it is promotion a culture of safety and security on chemical, biological, radiological and nuclear (CBRN) domains at regional level. The main interest of the CoE Initiative is building and/or enhancing national capacity to address CBRN threats in sensitive parts of the world themselves (mainly in five regions: African Atlantic Façade, Middle East, North Africa, South East Africa and South East Europe, Caucasus, Moldavia, Ukraine) in order to build a stronger EU security infrastructure.

The EU declared that supports and strengthens the BWC "national implementation measures, including administrative, judicial and criminal legislation, control over pathogenic microorganisms and toxins, (…) adoption of appropriate standards on biosafety and biosecurity measures (…) and development of national regulatory frameworks on biosafety and biosecurity [56]."

12.4.3 National Implementation

Based on the BWC provisions each State Party should "implement measures to minimise risks focus their national legislation, regulations and standards on safeguarding the workforce handling biological materials and on the protection of the environment, including the population, against accidental release or loss of hazardous materials. After 2011, (...) some State Parties focus their approaches on the physical protection of biological weapons-related biological materials to prevent unauthorised access by theft or diversion by non-State actors, including terrorists" [57]. An authoritative institutions or organizations may provide a support for states, how to create or enhance national biosafety and biosecurity system.

The Verification Research, Training and Information Centre (VERTIC) is an independent non-governmental organization (NGO), which supports State Parties in development, implementation and verification of international agreements related to disarmament, non-proliferation of biological and chemical weapons, and dual-use risks associated to biotechnology industry. At the Workshop on the Implementation of the Biological Weapons Convention, which was held in 3 August 2015 at Geneva, Mr. Scott Spence had presented a lecture on national implementation measures. He is a Director for National Implementation Programme and according to him the national implementation measures should include (1) definitions, (2) prohibitions and penalties, (3) jurisdiction, (4) biosafety and biosecurity, (5) transfer control and (6) enforcement issues. Regarding biological safety and security measures, he said that biosafety measures should mostly focus on prevention of unintentional exposure to pathogens and toxins or their accidental release. In contrast to biosafety, biosecurity measures need to concentrate on prevention of unauthorized access, loss, theft, misuse, diversion or international release of biological agents and toxins. Moreover, some specific biosafety and biosecurity procedures should be also established, specifically, a list of controlled biological agents and toxins, national system of notification about accidents, loss or theft, comprehensive record-keeping, physically secure of facilities, professional training on biosafety and biosecurity topics for personnel in addition to secure transportation for biological agents and dangerous goods [58].

Countries may choose different means of implementing internationally agreed principles through binding and/or non-binding national instruments. For this reason, there is a wide range of solutions that may be adopted at national level, including a variety of schemes, frameworks and instruments addressing to biosafety and biosecurity issues [59].

In the 7th Annual International Symposium – Biosecurity and Biosafety: Future Trends and Solutions, which was held in March 2016 in Italy, Mr. Scott Spence mentioned the latest Report on National Implementing Legislation [60]. According to the data included in the report many State Parties still need to strengthen their legal framework to ensure effective surveillance of activities related to hazardous biological agents and toxins. Many of countries involved at this Report should enhance biological safety and security by creation of independent supervisors, develop procedures and policies for authorizing certain studies and related publications, meet the challenges of the biological weapons prohibitions posed by increasing availability of dual-uses agents, toxins, equipment and technology, and present biosafety and biological culture in all relevant communities by dedicated codes of conduct [61].

Apart from above, also other international organizations made efforts to present their suggestions regarding legislative or policy reforms in context to biosafety and biosecurity on domestic level. One of them is the United Nations Environmental Protection Biosafety Toolkit, according to which biosafety and biosecurity legislation should comprise (1) biosafety policy providing an overarching framework and clear principles; (2) a regulatory regime; (3) means to address notifications or requests for authorizations; (4) means for enforcement and monitoring; and (5) public information, education and participation mechanisms [62].

The safety and security systems against biological threats will be effective only, if will be implemented and fully respected at all levels of government – domestic law, regional agreements, general principles and international treaties.

12.5 Non-legally Binding Instruments

Owing to the fact that law is inherently conservative and usually works with lags to bioscience, which is accelerative, aggressively changed and rapidly evolving there is a real danger that law just cannot keep up. Moreover, with the passage of time, gaps between scientific risks and legal control may significantly widen.

As mentioned above, legally agreements and legislative measures provide an overall rules binding for each State Party. In contrary to them, non-binding instruments offer an advantage of being faster and simpler to adopt than binding agreements, and provide more flexible means for update and amendment. Moreover, in necessity, it is possible to put them directly into practice at once. This kind of opportunity offers standards, guidelines, manuals, codes of conduct, good practices, etc., which are usually published by specialized organizations dealing with narrow area of particular activity. Moreover, many international norms and regional legislations are based on existing biosafety and biosecurity standards. They are being developed with help of international organizations or NGOs dealing with dedicated issues directly, e.g. WHO – on global public health security, Food and Agriculture Organization (FAO) – in achieving food security, or World Organization for Animal Health (OIE) responsible for improving animal health worldwide, as well as authorizing institutions like European Committee for Standardization (CEN) and International Organization for Standardization (ISO). Some actions are conducted with collaboration, e.g. Global Health Security Agenda (GHSA) [63]. Other professional bodies also provide publications [64, 65] and freely available online resources in this field [66].

Specialized associations likeABSA International - The Association for Biosafety and Biosecurity promotes biosafety as a scientific discipline among biosafety professionals, students and post docs, public health and hospital workers, emergency responders and scientists via development of professional standards, guidelines and regulations, providing training courses, webinars, conferences and workshops relevant to safe work, practices, lab equipment and instructional design of facilities [67]. The ABSA had been also active in biosecurity area for several years [68]. In similar way works the Asia-Pacific Biosafety Association (A-PBA) [69], the European Biological Safety Association (EBSA) [70] and the International Federation of Biosafety Associations (IFBA) [71].

Standards specified for biosafety and biosecurity management are dedicated for both, governmental and private institutions and following, are a helpful tool for State Parties for more effectively implementation of the international obligations in this field.

12.5.1 Biological Risk Management

Bearing in mind that the most probably exposition to potentially hazardous biological agents occurs in workplaces dealing with microbial agents directly, the WHO published some manuals strictly dedicated to laboratory personnel and health-care workers. One of them is a practical guidance on biosafety techniques for use in laboratories at all levels. The "Laboratory biosafety manual 3rd edition [72]" includes technical information on relation of risk group classification to biosafety levels (basic – Biosafety Level 1, basic – Biosafety Level 2, containment – Biosafety Level 3 and maximum containment – Biosafety Level 4) as well as laboratory type, laboratory practice and laboratory safety equipment. The Manual provides also types of personnel protective equipment, information of waste handling, accident reporting, laboratory animal handling, Standard Operational Procedures (SOPs), Good Laboratory Practice (GLP), safe collection and transportation of specimens for laboratory testing, secure transport of particularly dangerous pathogens, biosecurity activities in laboratories, codes of behavior, code of ethics for scientific research or codes of bioethics and so on. Besides, "Biorisk management: Laboratory Biosecurity Guidance [3]" consists of many specific aspects dealing with biological risk management and countering biorisk in biosecurity context, i.e. focusing on prevention of unauthorized access, theft, misuse, diversion or intentional release of agents and toxins from lab facilities.

To strengthening human health security by implementing the IHRs, WHO provides also other documents [73–75] and training courses associated to biorisk management (how to identify and control biosafety and biosecurity risks in laboratories, make a risk assessment or plan mitigation of risk, transport of infectious substances) [76]. Providing recommendations to specific present-day diseases or outbreaks also advance public safety and security [77, 78].

To set up requirements necessary to control risks associated with handling, storage and disposal of biological agents and toxins in laboratories and facilities almost

30 European countries agreed that an effective management system should be built. The concept of this system is based on a cycle of planning, implementing, checking, reviewing and improving performance and control of biorisk, which should be frequent upgraded. In result, the "Laboratory biorisk management standard [79]" presenting above concept was published as an effect of the CEN Workshop Agreement [80]. One more, the CWA 16335 describes biosafety and occupational measures as a professional competences for biosafety professionals, managers and trainers.

Likewise, the Office of Safety, Health and Environment of the US Centers for Disease Control and Prevention (CDC) in addition to the U.S. Department of Health and Human Services, the Public Health Service (PHS) and the National Institutes of Health (NIH) periodically update information and publish guidelines related to biosafety and biosecurity among workers in biological and medical laboratories [81]. Website promoting communication, transparency and awareness about biosafety, biocontainment, and laboratory biosecurity issues may be a good inspiration for finding the balance between Science, Safety and Security (S3), too [82].

12.5.2 Animal Safety and Security

Most of emerging infectious diseases are zoonotic and circulate in animal reservoirs before they cross over to infect humans. The response to all outbreaks of infectious disease is the same whether it is directed against naturally occurring infection, deliberate or accidental release. Many animal pathogens may be used as bioweapon because they have a high impact of infectivity, are cheap, easy to acquire, proliferate and unnoticed smuggled through border control. Moreover, molecular engineering of them may increase their virulence or make them more difficult to combat. In case of zoonotic disease coordination of both, animal health and public health officers is essential for quick diagnosis identification and confirmation, surveillance and the right trade of animals or products of animal origin. Control measures are often focused on eliminating a pathogen in the animal source.

The World Organization for Animal Health (OIE) collaborates in strengthening global biological security with many regional authorities and bodies. The OIE promotes reduction of health threats posed by animal-borne diseases through building a global surveillance and intelligence system for animal diseases and zoonotic situations, worldwide cooperation, policies, maintaining expertise and recommendations. It develops standards and guidelines to supports its Member Countries in protection themselves against transmission of diseases or pathogens during trade animals and animal products while avoiding unjustified sanitary barriers. The main objective of the OIE's standards is to recommend actions that will ensure biosafety and biosecurity via prevention of pathogenic biological agents transmission to animals, humans and environment [83].

In order to ensure adequate level of protection biosafety and biocontainment measures for veterinary laboratory workers handling hazardous biological materials should be put into practice. A manuals of diagnostic tests and vaccines developed in collaboration with the WHO, provides not only a guidelines for establishing of laboratory biosafety, but also biocontainment measures focusing on prevention accidental or deliberate release of pathogens into the environment [84, 85]. The OIE contributes to ensure the safe and secure processes in facilities operational infectious animal agents thru setting out the management and technical competence for the testing accreditation zoonotic diseases, as well [86]. Further, the OIE develops and prints a codes, guides and manuals to help States detect and prevent spread of aquatic and terrestrial animal illness, including diseases related to biological weapons [87, 88].

Additionally, the International Veterinary Biosafety Workgroup (IVBWG) dealing with biosafety issues in high containment (BSL 3 and above) animal facilities had also developed a very useful handbook in this field [89].

12.5.3 Food Safety and Security

From biological safety and security point of view, the Food and Agriculture Organization of the United Nations (FAO) supports countries in protection of food security, as well as prevention and control food safety risks. The FAO had developed and published number of guidelines dedicated to actual microbiological threats in food. For instance, the "Biosafety Resource Book [90]," which is a result of the FAO's biosafety capacity development training courses organized from 2002 to 2010. The book provides biosafety regulators, policy-makers and members of national biosafety committees with reference materials that can be readily consulted beyond the training events, when the need arises.

The another one is the "Biotechnology for Agricultural Development [91]," which contains the proceedings of the FAO international technical conference dedicated to the Agricultural Biotechnologies in Developing Countries that was held on 1–4 March 2010 in Mexico. The major objective of this conference was to take stock of the application of biotechnologies across the different food and agricultural sectors in developing countries in order to learn from the past and to identify options for the future to face the challenges of food insecurity, climate change and environmental degradation.

In order to effective respond to food safety threats and reduce a food-borne risk, multiagency cooperation and state emergency response actions need to be launched. The FAO with collaboration with the WHO had identified and developed a framework for food safety emergency response plan ready for approval at national level [92]. The "FAO/WHO guide for application of risk analysis principles and procedure during food safety emergencies [93]" presents continuation and expansion of the aspect how to strengthening resilience to food safety emergencies at the country level. The next one assists countries in establishing and implementation of an effective national system for food safety emergencies [94].

12.5.4 Biodiversity Protection

Biodiversity or biological diversity is a variety of all living species on earth. It is include plants, animals, microorganisms, their genes as well as terrestrial, marine and freshwater ecosystems of their lives. Biodiversity is an important building element for many human goods and services, because is fundamental not only to our health, (clean air, fresh water, food products), but also provides products making our life easier.

The associations between biodiversity, biosecurity and biosafety are as complex as the ecosystems they create or protect. Biosafety generally is used to describe frameworks of policy, regulation and management to control potential risks associated with the use of new biotechnologies. Biosafety instruments address the risks posed to the environment and human health when LMOs or GMOs are released into the environment either for research (e.g. small-scale or field-testing) or for commercial purposes. Biosafety instruments and food safety instruments address, respectively, contained use of GMOs and the risks posed to humans by genetically modified foods.

According to the FAO, biosecurity encompasses policy and regulatory frameworks to manage risks associated with agriculture and food production. This includes introduction and release of LMOs and GMOs and their derived products, introduction and spread of invasive alien species, alien genotypes and plant pests, animal pests, diseases and zoonosis [95]. Similarly, according to Mrs. Lois Ransom from the Australian Government Department of Agriculture, Fisheries and Forestry, biosecurity in biological diversity context is defined as management of risks to the economy, the environment and the community of pests and diseases entering, emerging, establishing or spreading [96].

Various international, regional and professional organizations deal with biological safety and security in this field. For example, UN Environment Programme – Global Environment Facility Initial Strategy on Biosafety (UNEP-GEF) [97] assists in the development of National Biosafety Frameworks through information sharing, collaboration and capacity building initiatives at the regional and sub-regional level. Numerous training workshops and materials in this fields are available at UNEP-GEF Biosafety Projects web page [98]. Also, Organization for Economic Co-operation and Development (OECD) [99] covers biotechnology policies, bioeconomy, biosafety, intellectual property rights and a research programme on biological resources in agriculture areas. The both, the "OECD Best Practice Guidelines for Biological Resource Centres (BRCs) [100]" and the "OECD Best Practice Guidelines on Biosecurity for BRCs [101]" meet biosafety and biosecurity issues.

12.5.5 Transport

Owing to the fact that infectious diseases do not respect borders, biological agents represent a significant potential threat to public health. To limiting the spread of infectious diseases worldwide common standards for safe and secure transportation of biological agents need to be used in practice, as well. Lack of adequate border control may result social and economic consequences for states, regions or worldwide. In reference to this, WHO had prepared and published a guide on transport of infectious substances [102]. It provides information for classifying infectious substances for transportation by all modes of transport and ensuring their safe packaging based upon levels of risk. Moreover, in order to provide protection and expeditious transport of these materials a good communication between the sender, the carrier and the receiver should be maintained. This guidance is regularly updated including the changes that apply each year.

Moreover, transport of dangerous goods related to nuclear, chemical and biological weapons by road, rail, inland, waterway and air is covered by numerous guides developed by many organizations specialized in transportation [103–107]. Some of them provide also trainings and publications available on their web pages [108].

Today, many of guidelines for safety and secure transport of dangerous goods had been translated into laws or binding regulations and put into practice [109–112].

12.5.6 Dual-Use Nature of Biological Researches

The possibility of using bioscientific knowledge for peaceful or malicious intention shows the dual-use dilemma [113] and affects both, publication content and biological materials themselves. The obvious point is that in order to learn how to defeat any disease we need to learn how it mechanism works. Thus, at the core of research dedicated for protection against biological violence is a paradox because methods that generate life-saving improvement (vaccines, treatments, etc.) are the same methods that could generate catastrophic damage. Some say that biosecurity is balancing the promise of biotechnology with preventing the biological weapons threats [18] and the efforts to strengthen biosecurity and reduce the threats posed by biological weapons require moral and ethical responsibilities of life scientists.

Obviously is fact that any crime should be punishment by law, nevertheless, in order to do it we need to specify what kind of scientific activities – methods or techniques should be forbidden. It is impossible if we use the same techniques and reagents for both. Moreover, there is lack of devices would detect unlawful take biological agents out of lab facility. Therefore, only thru building individual responsibility of themselves and our nature we can stimulate a conscience of scientists to consider pro and con of their achievement's application. We can raise their awareness on biosafety and biosecurity issues though education scientific community about the nature of the dual-use methods in biotechnology, reviewing plans for experiments or detailed methodology of sensitive publication, ensuring a role of life sciences in efforts to prevent bioterrorism and biowarfare.

Ethical codes and codes of conduct are primarily awareness-raising tools to remind scientists that they should contemplate potential consequences of their research. Such codes can also play a role in undergraduate and postgraduate education as part of education program in order to prepare students to think consequences of their activities, including any likely side effects. The content of codes of conducts usually involve fundamental issues such as awareness, safety and security, education and information, accountability and oversight. By biological safety and security should be understood the fact that scientists working with infectious agents such as natural or synthetic pathogens or dangerous toxins must be responsibly for use good, safe and secure laboratory procedures, whether codified by law or common practice [114]. They could also provide frameworks or guidance how develop a response to particular discoveries.

The *Biosecurity Working Group* under the *InterAcademy Panel* (IAP, The Global Network of Science Academies) [115] is a global network of the world's science academies, whose goal is to join forces to advise citizens and public officials on the scientific aspects of critical global issues. The IAP bringing together biosecurity experts from many countries in the world (inter alia USA, Canada, China, Australia, Poland, Nigeria, Cuba, Egypt, United Kingdom and Russia) also promote good biosecurity practices among scientists. The IAP Biosecurity Working Group is a regular participant of both, the BWC Meeting of Experts and the BWC Meeting of State Parties, where presents actual implications of scientific research for the BWC and biosecurity [116]. Most of ethical codes were developed and published aftermath discussions the IAP Biosecurity Working Group with close cooperation of scientific and international community [117].

Moreover, during the Eighth Review Conference of the BWC, European Union underlines the risks and threats posed by the rapid advances in biological science, including the possible acquisition and development of a biological weapon by a terrorist group [118]. Also the two intersessional meetings (2003–2005 [119] and 2007–2010) had engaged academic research communities and industry, partly in an effort to explore biological threat reduction options through discussions and consultation on biosafety, biosecurity, disease surveillance, response and other policy and procedural mechanisms.

Apart from that frequently are organized symposiums, conferences or workshops during which different approaches and perspectives of how to regulate subjects related to the Dual Use Research of Concern (DURC) are deliberated, mainly focusing on microbes and raising connections to them concerning on biosafety and biosecurity issues or potential limits to the freedom of research [120].

12.5.7 Bioterrorism

Lots of natural occurring pathogens or toxins could be used for warfare or terrorism. The idea of using microorganisms as warfare agents has a very long history [121–124]. Over the years many international authorities dealing with counter threats of biological warfare came to the common conclusion that (1) more intensive cooperation for threat assessment and planning as well as (2) implementation of national legislation in line with the BWC may significantly reduce this kind of threat.

In development of the bio-preparedness plan or strategy should be involved not only policy makers, law enforcement, medical doctors or customs officers, but also any other relevant national agency representatives working in the area of bioterrorism, who will be or potentially may be engaged in prevention, protection, first response capacity, prosecution of criminals or terrorists, surveillance, research capacity, response and recovery aspects [125]. Cooperation among different sectors represented not only national, but also regional or international level may allow to develop a coordinated national prevention and response plan dedicated for biological terrorism.

The CDC had developed a system to prioritize biological agents according to their risk to national security and categorize them. The Category A agents includes the highest priority bioterrorism agents such as *Bacillus anthracis, Clostridium botulinum* toxin, *Yersinia pestis, Francisella tularensis* as well as variola major, filoviruses and arenaviruses. Following, the Category B agents are moderately easy to disseminate and result in low mortality. These include *Brucella* sp., *Burkholderia mallei, Coxinella burnetii, Rickettsia prowazekii* and other agents. And finally, the Category C agents involve emerging disease agents that could be engineered for mass dissemination in future, such as Nipah and hantavirus [126].

Strengthening public-health care and infectious disease medical infrastructure, adequate epidemiologic and laboratory capacity with high biosafety lab levels, rapid detection, identification and characterization of the agents, effective disease surveillance, emergency distribution of antibiotics and vaccines, secure collection of traces at the crime scene, border control, training of multi-agency response and communication systems during simulation of similar events are some of steps contribute for biodefence capability-building [127].

Furthermore, deliberate contamination of food is also a real and still actual threat, which may have global public health implications. Member States of WHO had expressed concern that chemical, biological or radio-nuclear agents might be introduced into food and other media to intentionally harm people many times. That is why, WHO had published some guidelines with technical information how to prevent and respond to intentional contamination of food. The first one "The Public health response to biological and chemical weapons: WHO guidance [128]" had been revised and published in 2004. This second edition of WHO's 1970 publication "Health aspects of biological and chemical weapons [129]" includes information designed to guide preparedness for and response to the deliberate use of biological and chemical agents that affect health.

Secondly, after 2008, WHO issued an update to the 2003 original guidelines "Terrorist Threats to Food – Guidelines for Establishing and Strengthening Prevention and Response Systems". This book responds to increasing concern in WHO's Member States that chemical, biological or radio-nuclear agents might be used deliberately to harm civilian populations and that food might be a vehicle for disseminating such agents [130].

We have a lot of various specialized agencies and international organizations dealing with bioterrorism issues directly or indirectly, beginning from prevalence public health policy, intelligence or secret service, law enforcement and public health bodies, finally to agencies responsible for resilience and return to the state before the event. Many of them are working on international or regional level – European Union (EU), North Atlantic Treaty Organization (NATO), Organization for Security and Co-operation in Europe (OSCE), International Committee of the Red Cross (ICRC), United Nations Interregional Crime and Justice Research Institute (UNICRI), United Nations Institute for Disarmament Research (UNIDIR), United Nations Office for Disarmament Affairs (UNODA), United Nations Security Council Resolution 1540 (UNSCR 1540) Group of Experts and provide guidelines, support or technical assistance to local and state agencies in developing coordinated preparedness plans and response protocols. They often organize conferences, workshops, trainings and provide self-assessment tools for terrorism preparedness including performance standards and/or attack simulations. In addition, they support applied research to develop innovative tools and strategies to prevent or mitigate illness and injury caused by biological terrorism.

Apart from that many NGOs like Biosecu.re, BioWeapons Prevention Project (BWPP), Center for Security Studies (CSS), Geneva Centre for Security Policy (GCSP), Geneva Disarmament Platform (GDP), Institute for Security Studies (ISS), International Office for Innovation in Reducing Crime Ltd. (IOIRC), Johns Hopkins Center for Health Security (CHS), World Anti-Bioterrorism Organization (WABO) deal with countering the threat posed by terrorist use of CBRN materials, too.

Legislation has a central role in countermeasures dedicated bioterrorism and proliferation of biological weapon. National legislature based on international agreements should (1) punish the production, stockpiling, transfer and use of bioweapons, (2) monitor the usage of biological agents and dual-use technology that lend development of bioweapons. Moreover, (3) establishment of national and international databanks that monitor transfer of biological agents and dual-use goods, their use in industry outreach programs, their licensed availability in national, regional and international markets as well as (4) establishment and use of confirmatory protocols in the destruction and dispersal of outdated stockpiles are recommended to mitigate the bioterrorism risk [131].

Additionally, BWC's Confidence-Building Measures (CBMs) sustained by use of monitoring and verification protocols contributes biosecurity measures via reducing and eliminating threats of biological warfare and bioterrorism. Also development of national guidance or setting of recommendations for biosecurity facilities working with, storing and transporting "high consequence pathogens", namely, in clinical laboratories, hospitals, research universities, private industry or numerous state and federal facilities will be also favorable.

Finally, without domestic legislative framework, any native agency or emergency service cannot respond to or initiate investigation of any crime [132].

12.6 Conclusion

Each incident with high risk biological agents, no matter about intention, poses a real danger for human, animals, plants and ecosystems they live. Biological agents acquired from natural sources or genetic engineering labs may cause severe illnesses or deaths, unless adequate preventive measures are used. Application of biosafety measures by employees will minimalize a risks associated with biological agents in the workplace. On the other hand, an effective biological threats assessment and management using biosecurity measures may prevent or reduce probability of deliberate release.

Based on lessons learned from history several preventive strategies against terrorism using weapons of biological warfare had been developed. Due to wide scope of biosafety and biosecurity interests synchronized strategy of different institution dealing with biological agents should be discussed. One of the most important is implementation of legislative framework and adherence a rigorous policy in reference to biological safety and security. The both, legally and non-legally binding instruments on biosafety and biosecurity ought to be implemented on each governmental level – international, regional and domestic, respected and used in practice.

References

1. Dictionary of Military and Associated Terms. US Department of Defense 2005
2. BWC/MSP/2005/MX/WP.14, Infectious Diseases, Biosafety and Biosecurity – Prepared by Germany, 13 June 2005
3. WHO Biorisk Management, Laboratory biosecurity guidance, 2006
4. WHO Laboratory biosafety mannual, 3rd ed, 2004
5. OECD – Glossary of Terms. http://www.biosecuritycodes.org/gloss.htm
6. Strengthening a Global Biosecurity/Biosafety Framework and Coping with the Biotechnology Revolution, BWC Implementation Support Unit
7. WHO, Biorisk Management: Laboratory Biosecurity Guidance, September 2006
8. Glossary of the FAO Basic Laboratory Manual for the Small-Scale Production and Testing of I-2 Newcastle Disease Vaccin
9. Agenda item 5 of Annex 1: national, regional and international measures to improve biosafety and biosecurity, including laboratory safety and security of pathogens and toxins, BWC/MSP/2008/MX/3, Report of The Meeting Of Experts, 8 September 2008
10. BWC/MSP/2008/MX/INF.1 Biosafety and Biosecurity – Submitted by the Implementation Support Unit, 24 June 2008
11. BWC Convention, Article I
12. BWC Convention, Article II

13. BWC Convention, Article III
14. BWC Convention, Article IV
15. BWC/CONF.VI/6, Final document of sixth review conference, 20 Nov–8 Dec 2006, Geneva
16. BWC Convention, Article V–VI
17. BWC Convention, Article VII
18. http://www.nti.org/about/biosecurity/
19. BWC Convention, Article X
20. The Cartagena protocol on biosafety to the convention on biological diversity, Article I
21. UN Security Council Resolution 1540, OP1–3
22. WHO International Health Regulations, 2005
23. Directive 2000/54/EC of the European Parliament and of the Council of 18 September 2000 on the protection of workers from risks related to exposure to biological agents at work (seventh individual directive within the meaning of Article 16(1) of Directive 89/391/EEC)
24. Directive 2009/41/EC of the European Parliament and of the Council of 6 May 2009 on the contained use of genetically modified micro-organisms
25. Directive 2001/18/EC of the European Parliament and of the Council of 12 March 2001 on the deliberate release into the environment of genetically modified organisms and repealing Council Directive 90/220/EEC
26. Directive (EU) 2015/412 of the European Parliament and of the Council of 11 March 2015 amending Directive 2001/18/EC as regards the possibility for the Member States to restrict or prohibit the cultivation of genetically modified organisms (GMOs) in their territory
27. Regulation (EC) No 1829/2003 of the European Parliament and of the Council of 22 September 2003 on genetically modified food and feed
28. Regulation (EC) No 1830/2003 of the European Parliament and of the Council of 22 September 2003 concerning the traceability and labelling of genetically modified organisms and the traceability of food and feed products produced from genetically modified organisms and amending Directive 2001/18/EC
29. Regulation (EC) No 1946/2003 of the European Parliament and of the Council of 15 July 2003 on transboundary movements of genetically modified organisms
30. Directive 2009/147/EC of the European Parliament and of the Council of 30 November 2009 on the conservation of wild birds
31. Council Directive 92/43/EEC of 21 May 1992 on the conservation of natural habitats and of wild fauna and flora
32. Communication from the Commission to the European Parliament, the Council, the Economic and Social Committee and the Committee of the Regions: our life insurance, our natural capital: an EU Biodiversity Strategy to 2020 (COM(2011) 244)
33. Communication from the Commission to the European Parliament, the Council, the Economic and Social Committee and the Committee of the Regions: a new EU Forest Strategy: for forests and the forest-based sector
34. Convention on International Trade in Endangered Species of Wild Fauna and Flora, 1975
35. European Agreement concerning the International Carriage of Dangerous Goods by Road (ADR)
36. Regulations concerning the International Carriage of Dangerous Goods by Rail (RID, part of the Convention concerning International Carriage by Rail)
37. IATA Dangerous Goods Regulations 57th ed (DGR), Jan 2016
38. IATA Dangerous Goods Regulations 53rd ed, Guidance Document. Infectious substances
39. International Maritime Dangerous Goods Code (IMDG Code, part of the International Convention for the Safety of Life at Sea)
40. Directive 2008/68/EC of the European Parliament and of the Council of 24 September 2008 on the inland transport of dangerous goods
41. Directive 2008/54/EC of the European Parliament and of the Council of 17 June 2008 amending Council Directive 95/50/EC on uniform procedures for checks on the transport of dangerous goods by road, as regards the implementing powers conferred on the Commission

42. Commission Directive 2004/110/EC of 9 December 2004 adapting for the sixth time to technical progress Council Directive 96/49/EC on the approximation of the laws of the Member States with regard to the transport of dangerous goods by rail
43. Directive 2010/35/EU of the European Parliament and of the Council of 16 June 2010 on transportable pressure equipment and repealing Council Directives 76/767/EEC, 84/525/EEC, 84/526/EEC, 84/527/EEC and 1999/36/EC
44. Council Regulation (EC) No 428/2009 of 5 May 2009 setting up a Community regime for the control of exports, transfer, brokering and transit of dual-use items
45. BWC/CONF.VI/WP.22 Bioterrorism – submitted by Italy on behalf of the European Union, 21 Nov 2006
46. EU Strategy against Proliferation of Weapons of Mass Destruction (WMD), 2003
47. Fight against the proliferation of weapons of mass destruction – EU strategy against proliferation of Weapons of Mass Destruction (WMD), Brussels, Dec 2003
48. Council Joint Action 2008/307/CFSP of 14 April 2008 in support of World Health Organization activities in the area of laboratory bio-safety and bio-security in the framework of the European Union Strategy against the proliferation of Weapons of Mass Destruction
49. A green paper on bio-preparedness, 2007
50. Communication from the Commission to the European Parliament and the Council on Strengthening Chemical, Biological, Radiological and Nuclear Security in the European Union – an EU CBRN Action Plan
51. The European Union Counter-Terrorism Strategy
52. Regulation (EC) No 1717/2006 of the European Parliament and of the Council of 15 November 2006 establishing an Instrument for Stability
53. Science and Technology Center in Ukraine, Annual Report 2011, Ukraine, 2012
54. Regulation (EU) No 230/2014 of the European Parliament and of the Council of 11 March 2014 establishing the Instrument contributing to Stability and Peace
55. http://www.unicri.it/topics/cbrn/coe/
56. BWC/CONF.VIII/PC/WP.5 Position of the European Union relating to the Eighth Review Conference of the BWC, 12 Apr 2016
57. BWC/CONF.VI/WP.2 Biosafety and Biosecurity – submitted by Germany on behalf of the European Union, 20 Oct 2006
58. Workshop on the Implementation of the Biological Weapons Convention, Geneva, 3 Aug 2015
59. Bielecka A, Mohammadi AA (2014) State-of-the-art in biosafety and biosecurity in European countries. Arch Immunol Ther Exp 62(3):169–178
60. Biological Weapon Convention. Report on National Implementing Legislation, VERTIC National Implementation Measures Programme, Switzerland, 2016
61. Spence S., 7th annual international symposium – biosecurity and biosafety: future trends and solutions, Milan, Italy, 22–24 March, 2017
62. http://new.unep.org/tools/default.asp?ct=biosafe
63. https://www.ghsagenda.org/
64. Guidelines for Biosafety Laboratory Competency, Morbidity and Mortality Weekly Report, vol 60, CDC, April 2011
65. Guidelines for Safe Work Practices in Human and Animal Medical Diagnostic Laboratories, Morbidity and Mortality Weekly Report, vol 61, CDC, January 2012
66. https://www.aphl.org/programs/preparedness/Biosafety-and-Biosecurity/Pages/BB-Resources.aspx
67. https://absa.org/
68. ABSA Task Force on Biosecurity White Paper on Understanding Biosecurity, 2003
69. http://www.a-pba.org/
70. http://www.ebsaweb.eu/
71. http://www.internationalbiosafety.org/index.php/resources/biosafety-biosecurity/biosafety-guidelines

72. WHO Laboratory biosafety manual 3rd ed, 2004
73. Extended Biosafety Advisory Group meeting report Geneva, Switzerland, 24–26 Nov 2014, WHO, 2015
74. Laboratory Biorisk Management: Strategic Framework for Action 2012–2016, WHO, 2012
75. Responsible life sciences research for global health security, WHO, 2010
76. http://www.who.int/ihr/training/biorisk_management/en/
77. http://www.who.int/ihr/publications/bioriskmanagement_3/en/
78. https://absa.org/ebola/
79. CWA 16335 – Biosafety professional competence, Sept 2011
80. CWA 15793 – Laboratory biorisk management standard, CEN Workshop Agreement, 2011
81. Biosafety in Microbiological and Biomedical Laboratories, 5th ed, US HHS, PHS, CDC, NIH, 2009
82. https://www.phe.gov/s3/Pages/default.aspx
83. OIE Biological Threat Reduction Strategy – Strengthening Global Biological Security, 2012
84. OIE Manual of Diagnostic Tests and Vaccine Production, 2011
85. OIE Manual of Diagnostic Test and Vaccines and Terrestrial Animals, 2016
86. OIE Quality Standard & Guidelines for Veterinary Laboratories, 2008
87. OIE Terrestrial Animal Health Code, 2016
88. OIE Aquatic Animal Health Code, 2016
89. Veterinary Containment Facilities: Design and Construction handbook, IVBWG, 2006
90. FAO Biosafety Resource Book, 2011
91. FAO Biotechnology for Agricultural Development, 2011
92. FAO/WHO framework for developing national food safety emergency response plans, 2010
93. FAO/WHO guide for application of risk analysis principles and procedure during food safety emergencies, 2011
94. FAO/WHO guide for developing and improving national food recall systems, 2012
95. http://www.fao.org/biodiversity/cross-sectoral-issues/biosecurity/en/
96. Biodiversity and World Food Security: Nourishing the Planet and Its People Conference, Ransom L., Preventing Biodiversity, Promoting Biosecurity and Biosafety: Australian Perspectives, 30 Aug–1 Sept, 2010, Canberra, Australia
97. http://www.unep.org/biosafety/
98. https://web.archive.org/web/20060824010149/http://www.unep.ch:80/biosafety/BCH.htm
99. http://www.oecd.org/
100. OECD Best Practice Guidelines for BRCs, 2007
101. OECD Best Practice Guidelines on Biosecurity for BRCs, 2007
102. WHO Guidance on regulations for the transport of infectious substances 2015–2016, 2015
103. The International Maritime Dangerous Goods (IMDG) Code, 2016
104. The International Ship and Port Facility Security (ISPS) Code
105. ICAO The Technical Instructions for the Safe Transport of Dangerous Goods by Air, 2017–2018 edition
106. IATA Guidelines for Transport of Infectious Substances by Air
107. UNESC Recommendations on the Transport of Dangerous Goods
108. http://www.iata.org/whatwedo/cargo/dgr/Pages/download.aspx
109. The Carriage of Dangerous Goods and Use of Transportable Pressure Equipment Regulations (CDG)
110. The European Agreement concerning the International Carriage of Dangerous Goods by Road (ADR)
111. IATA The Dangerous Goods Regulations
112. HSE Guide to the Control of Substances Hazardous to Health (COSHH) Regulations
113. Selgelid Michael J (2009) Governance of dual-use research: an ethical dilemma. Bulletin of the WHO 87:720–723
114. IAP Statement on Biosecurity, 2005
115. www.interacademy.net

116. Report of the Meeting of Experts of the BWC, 4–6 Aug 2014, Geneva, Switzerland
117. BWG A Code of Conduct for Biosecurity, 2007
118. Position of the European Union relating to the Eighth Review Conference of the BWC submitted by the European Union, 26–27 Apr and 8–12 Aug 2016, Geneva
119. WHO Life Science Research: Opportunities and Risks for Public Health. Mapping the Issues, 2005
120. "Dual Use Research on Microbes: Biosafety, Biosecurity, Responsibility" Symposium, 10–12 Dec 2014, Hannover, Germany
121. Alchon SA (2003) A pest in the land: new world epidemics in a global perspective. University of New Mexico Press, Albuquerque
122. Oldston M (2010) Viruses, plagues, & history. Oxford University Press, New York City
123. Carus W (2002) Bioterrorism and biocrimes: the illicit use of biological agents since 1900. Fredonia Books, Amsterdam
124. Thompson C (2006) The bioterrorism threat by non-state actors: hype or horror? Naval Postgraduate School, Monterey
125. EC Green Paper on Bio-Preparedness, 2007
126. Bioterrorism Agents/Diseases (by Category), Emergency Preparedness and Response, CDC
127. Hamburg, Margaret A (2002) Preparing for and preventing bioterrorism. Issues in Science and Technology 18(2) Winter
128. The Public health response to biological and chemical weapons: WHO guidance, 2004
129. WHO Health aspects of biological and chemical weapons, 2001
130. WHO Terrorist Threats to Food – Guidelines for Establishing and Strengthening Prevention and Response Systems, 2008
131. DaSilva Edgar J, (1999) Biological warfare, bioterrorism, biodefence and the biological and toxin weapons convention. EJB Electro J Biotechnol.2(3), Issue of Dec 15
132. Roffey R, Lantorp K, Tegnell A, Elgh F (2002) Biological weapons and bioterrorism preparedness: importance of public-health awareness and international cooperation. Clin Microbiol Infect 8(8):522–528

Chapter 13
The Role of Bioforensics in Medical Bio-Reconnaissance

Lothar Zöller and Gelimer H. Genzel

Abstract Since the 1990s, a broad spectrum of regional conflicts and crises have evolved that have been accompanied by a growing threat of international terrorism. How vulnerable our modern societies would be towards a covert biological attack became evident in the 2001 anthrax letters attack in the United States. Biothreats are currently associated with asymmetric warfare scenarios and non-state actors rather than with state-driven biowarfare facilities. Against this backdrop, NATO has to consider biological warfare and bioterrorism as a serious threat to its forces. In bioterroristic scenarios the deliberate release of a biological agent will most probably remain undetected until a cluster of cases will suggest an unusual outbreak of disease. In military settings it is primarily the responsibility of the Medical Services to recognize the outbreak and to launch an appropriate outbreak investigation. Major goals of a medical bio-reconnaissance mission are to rapidly identify the causative agent of the outbreak and to differentiate between natural and deliberate outbreaks. In contrast to the investigation of overt natural outbreaks, forensic aspects have to be considered and appropriate procedures have to be implemented quite from the beginning when unusual outbreaks are to be investigated. If a biothreat agent is detected, it may be necessary to enter further genetic analysis in order to differentiate between natural and intentional outbreaks and to trace back the origin of the agent. Microbial forensics is mainly concerned with taking molecular fingerprints of biothreat agents by means of molecular typing techniques enabling the investigator to identify and trace back a particular strain by comparing it with the fingerprints stored in a typing database. The bioforensic approach may well be capable of elucidating the source of an outbreak as has been evidenced in the Amerithrax case in 2001. In order to detect molecular differences of microbial strains, a number of sophisticated typing techniques are currently employed, the most recent of which is whole genome sequencing, which has even entered the field laboratories by means of portable next generation sequencing devices like the MinION™.

L. Zöller (✉) · G. H. Genzel
Bundeswehr Institute of Microbiology (BwIM), Munich, Germany
e-mail: lotharzoeller@bundeswehr.org

V. Radosavljevic et al. (eds.), *Defence Against Bioterrorism*, NATO Science for Peace and Security Series A: Chemistry and Biology,
https://doi.org/10.1007/978-94-024-1263-5_13

13.1 Approach to the Term Bioforensics

The term "bioforensics" or "microbial forensics" came into common use during the investigation of the anthrax letter attacks in the United States in 2001 although it had been mentioned already years before by FBI forensic experts. The investigational approach, which is also known as the "Amerithrax" investigation, was unique at that time in involving criminologists, namely the FBI, as well as microbiologists and genome sequencing experts [1, 2]. Their common goal was to trace the anthrax bacilli used in the attacks back to their source and to link it to the perpetrator, which overall took 9 years of investigation. This kind of trace-back analysis to solve a biocrime or a bioterroristic act gave name to a new scientific discipline called microbial forensics [3]. The approach is very similar to the bundle of methods that are nowadays often applied to trace the sources of natural infectious disease outbreaks, which is also known under the term "molecular epidemiology".

Generally spoken, bioforensics or forensic biology applies the biological science to the investigation of legal matters. The goal of bioforensics is to answer the question who committed the offense. In that logic microbial forensics is a subdiscipline of bioforensics dealing with microbes and their characteristics as evidence in a criminal investigation. It has turned out to be a helpful instrument in cases where microbes have been used as a weapon, but may also be useful in cases where microbes provide a characteristic signature on evidences, thus helping to trace their origin. But, during the anthrax letter attacks in 2001 the delay in the diagnosis of the first cases and the long duration of the molecular trace-back analysis, which finally led to the identification of the perpetrator, also revealed the gaps in responding to unusual biological events.

Even though the term bioforensics obviously describes a wide and complex scientific field, the official NATO Glossary is using the term "CBRN forensics" in a more general way to designate "the scientific methods and techniques used to analyze materials and data in support of a chemical, biological, radiological and nuclear incident or threat investigation" [4]. In a military setting bioforensics as a subdiscipline of CBRN forensics designates a complex of procedures and scientific methods that allow the discrimination between a biological attack and a natural outbreak and the attribution of a biological attack to the perpetrator. Bioforensics is part of bioreconnaissance, which in a wider sense designates all measures that aim to detect, recognize, trace back and attribute the deliberate release of a biothreat agent. Biothreats are currently associated with asymmetric warfare scenarios and non-state actors (bioterrorism) rather than with state-driven biowarfare facilities. In bioterroristic scenarios a covert release of the biological agent must be expected, which will most probably remain undetected until a cluster of cases will suggest an unusual outbreak of disease in the target population. This is why a biological attack will most probably manifest in a way that in the first instance will involve the medical facilities and, in a military setting, the Medical Services of the Armed Forces. Samples recovered from the casualties will not only be needed for a rapid and proper diagnosis but will also constitute pieces of evidence in an investigation that might result in serious legal, political or even military consequences. The Armed Forces Medical Services must therefore become aware of this new role they are playing in case of a biological attack and implement appropriate forensic procedures. Over the past 10 years the Medical Service of the German Armed Forces has built up a concept to implement bioforensic procedures in its bio-reconnaissance facilities, which is outlined in the following.

13.2 Principles of Bio-Reconnaissance

13.2.1 Natural vs. Intentional Outbreaks

It is highly unlikely that the deliberate release of a biological agent in the framework of a biological attack would be detected before the agent would reach the target population or infrastructure. Due to the variety of potential biological agents and associated routes of delivery as well as the need to distinguish them from non-human pathogens, current biomonitoring systems are far from providing reliable real-time detection that would allow to implement protective measures or even to warn the target population in due time [5]. Only in scenarios where a biothreat is recognized prior to the use of bioweapons, i.e. where an enemy threatens to use bioweapons or is known to possess them, aerosol detectors may be of value in providing indications of the release of airborne biological agents. This may allow to implement post-exposure prophylactic measures or early treatment during the incubation period (detect-to-treat capability).

An outbreak is basically defined as the occurrence of two or more cases of a disease or deaths that are closely related in epidemiological terms. Minor local outbreaks of an infectious disease often mark the initial phase of an epidemic. An outbreak can be characterized as "unusual" when its ecological, epidemiological, infectiological and microbiological characteristics are non-standard. "Standard" would be the typical occurrence of an infectious disease with the expected seasonal, geographic and demographic distribution patterns and the known clinical picture. However, the deliberate release of a biological agent must be suspected if the epidemiological, microbiological or clinical features of the disease are unknown, new, unexplainable or unexpected, for example if the pathogen is not known to be endemic in the area where the outbreak occurs. Atypical courses of disease that differ from commonly seen natural infection, e.g. inhalational anthrax or primary pneumonic plague, should also raise suspicion as should unusually high rates of manifestation and lethality. A non-natural cause of a disease may also be suspected in case of an unusually short incubation period or unknown or uncommon antibiotic resistance properties of an endemic pathogen. Intelligence or police evidence of a biological threat may provide additional clues. The intentional release of a biological agent would be even harder to reveal if a previously unknown or genetically engineered agent or a regional endemic pathogen were disseminated. In the latter case, the outbreak could even simulate a "natural" epidemic. Therefore, it is a basic need for the military Public Health facilities to continuously monitor the epidemiological situation of infectious diseases in the theatre of operation and to report anomalies strictly according to the requirements specified by each nation [6].

Biological warfare agents usually need hours to days to take effect after hitting the target population. Incubation period and clinical manifestations depend on the type, virulence, quantity and delivery of the agent as well as on the susceptibility of the exposed population (e.g. vaccination status, underlying diseases). The appearance of an outbreak with an unusual disease manifestation or an unusual epidemiologic pattern is, therefore, the most apparent clue to a covert biological attack. As soon as an outbreak has been recognized an outbreak investigation must be put in place, the first major goal of which is to identify the causative agent because in a proverbial sense "you cannot win the war if you don't know the enemy". The second major goal is to identify its potential source, e.g. food, water, animals or others. This is, by the way, equally important for natural as well for deliberate outbreaks.

If the outbreak is characterized as unusual, it will always be necessary to implement efforts to discriminate between a natural outbreak and an intentional event. A scoring system has been proposed that may give a first assessment and trigger further investigations [7, 8]. It will also be important in such a case to supplement the investigation with forensic procedures to preserve evidence.

13.2.2 Recognizing an Outbreak

After a covert biological attack an uncommon disease outbreak and/or sudden deaths will primarily alert rescue services, emergency physicians, resident physicians, internists, specialists in tropical and laboratory medicine, microbiologists and pathologists as well as doctors and nurses of outpatient clinics, emergency rooms and hospitals. Pharmacists may notice a sudden increase in antibiotic prescriptions. Epidemics in animal populations that may also be targets of biological attacks will first be recorded by animal owners and resident veterinarians and will be reported to the responsible animal health offices. Early detection of an outbreak largely depends on the awareness of physicians initially contacted by the patients and the public health facilities behind them whose responsibility is to recognize an outbreak and to put in place adequate epidemiological methods for outbreak investigation. In a military setting, physicians should, therefore, be routinely trained to recognize infectious disease outbreaks and to call in biodefence specialists as soon as the outbreak reveals "unusual" characteristics. In the Medical Service of the German Armed Forces this is a routine part of the postgraduate training of all physicians. A case definition must be developed early during the outbreak investigation in order to help first responders and physicians to recognize further cases and to provide a differential diagnosis. Because different biological agents may have been used

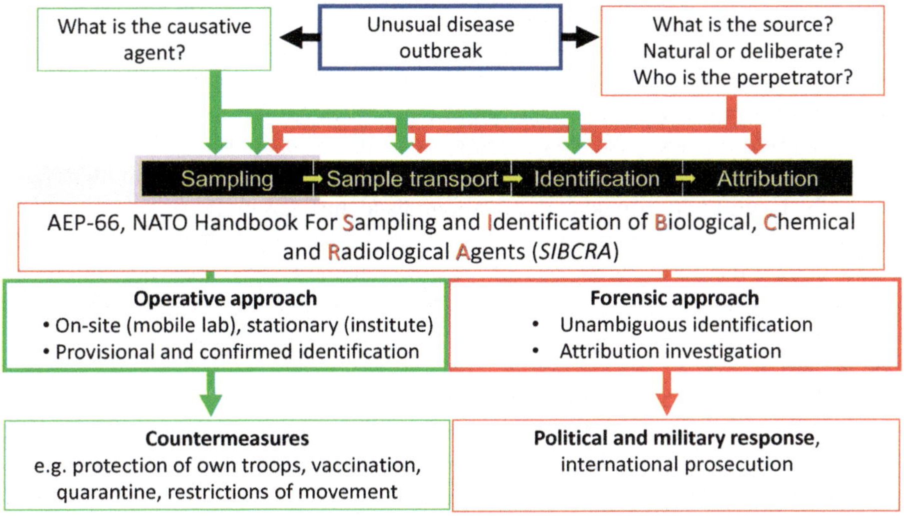

Fig. 13.1 Principles of bio-reconnaissance in the Medical Service of the German Armed Forces: operative and forensic approaches

simultaneously or at different times and in different ways (e.g. airborne, alimentary), the triage process must also consider different possible syndromes if applicable.

13.2.3 Outbreak Investigation in a Military Setting

13.2.3.1 Responsibility of the Medical Service

The bio-reconnaissance facilities of the Medical Service of the German Armed Forces comprise medical bio-reconnaissance teams who are trained to enter a scenario under adequate personal protective equipment, to assess it, and to recover the proper specimens. Standard operating procedures for the qualified recovery of samples from humans, animals, food and water and for appropriate sample transport under strict observance of forensic requirements have been established. The German medical bio-reconnaissance facilities also include methods for epidemiological investigation, e.g. to conduct a case control study in collaboration with military public health experts, and also methods for the unambiguous identification of the causative agent and for molecular trace-back analysis. The Medical Service of the German Armed Forces is, therefore, prepared to make decisive contributions to achieving the major goals mentioned in the previous section of this chapter.

In what is called an operative approach (Fig. 13.1), identification of the causative agent is primarily the responsibility of a field deployable laboratory, which is a modular system of devices, assays, protective equipment and trained staff that allows to perform diagnostic testing from specimens on-site at the level of provisional and in part confirmed identification. All the laboratory and personal protective equipment is packed in handy carrying cases and can be deployed by military or civilian aircraft within 48 h to any place where the action is needed [9]. The field lab is supported by the stationary reach back facilities of the Bundeswehr Institute of Microbiology (BwIM), which extends the level of confidence in agent identification to the levels of confirmed and unambiguous identification. Proper identification of the causative agent will always be a prerequisite for the implementation of adequate countermeasures such as vaccination, chemoprophylaxis, quarantine or restriction of movement. It is very time-sensitive, because the "window of opportunity", during which the outcome of the outbreak can be significantly influenced, is rather short.

In addition to this, in case of an unusual outbreak that might have been caused by the intentional release of a biothreat agent, a "forensic approach" (Fig. 13.1) must be implemented quite from the beginning and throughout all stages of the diagnostic process from sampling to identification.

13.2.3.2 Sampling

Sampling is a pivotal step in the first phase of an outbreak investigation. There might be only one chance to recover adequate samples from humans, animals (living or dead) or from the environment, which will be the basis for all consecutive laboratory examinations and the conclusions drawn from the results. This is especially important if the outbreak might have been caused by a biological attack. Forensic procedures must, therefore, be implemented quite from the beginning in order to preserve evidence and to safeguard the chain of custody. Insufficient procedures and flaws during the sampling process due to a lack of expertise cannot be compensated at a later stage of the investigation. Thorough photographic documentation of the sampling situation and the use of pre-controlled lots of sampling devices are mandatory. Samples should always be taken in duplicate. In order to transport the samples to a laboratory, sample vessels must be sealed by means of security labels wearing individual numbers. These labels display immediate evidence of attempted removal or manipulation. Furthermore, each sample must be wrapped in a special decontaminable transportation pouch. Thorough documentation of the sampling sites and samples must be done by setting up a record on tear-proof and autoclave-resistant paper. A major task of the sampling team is to safely and securely transport the samples to the lab by using appropriate transport vessels and boxes fulfilling the requirements of the international dangerous goods regulations. It is important to safeguard the chain of custody during the whole bio-reconnaissance process, which means that the transport chain of the samples must be fully supervised and documented from the sampling sites to the lab. To give an example of this, the deliverer of the specimens must be photographed with his identity card in hand (Fig. 13.2). Throughout the sampling, transport and subsequent laboratory analysis processes, the legal integrity of samples and data must be

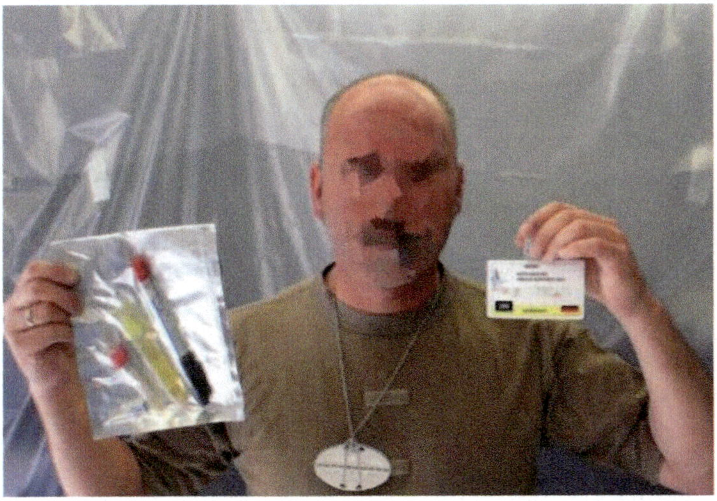

Fig. 13.2 Safeguarding the chain of custody: photo-documentation of sample delivery to the laboratory

guaranteed. Therefore, it is necessary to have an accurate written record to track the possession, handling, and location of samples and data from collection through reporting.

13.2.3.3 Mobile Laboratory Investigation

The German rapidly deployable laboratory facility is designed and equipped especially for the reconnaissance of intentional outbreaks. It is based on a modular concept of devices and assays that allow sampling and on-site diagnosis in various situations. The lab can be operated in multiple environments, even in a garage, but is also equipped with an inflatable lab environment to allow autonomous operation. Most of the diagnostic procedures are realtime-PCR-based and currently adapted to the SmartCycler™ (Cepheid Inc.) platform, but the lab also provides methods for light microscopy and antigen detection (Enzyme immunoassays, immunofluorescence assays, handheld testkits). Cultivation techniques are not provided at the level of the mobile lab. During the handling of potentially hazardous and forensically important samples in a mobile laboratory, appropriate biosafety precautions like the use of a low-pressure vented glovebox must be observed, not only to protect the laboratory staff from infection, but also to avoid any contamination that could later on lead to misinterpretation of the results. This is especially true for the sensitive step of DNA- or RNA-extraction. Retention samples should be saved from crucial steps in order to preserve the option to control doubtful results. Therefore, only highly trained and experienced laboratory and scientific personnel must operate such a diagnostic field laboratory.

13.2.3.4 Identification of the Agent

Arriving at the identification step, it is important to implement the highest possible level of confidence, which can be achieved by the combination of independent methods and independent diagnostic targets. The NATO SIBCRA Handbook [10] differentiates between the levels of provisional identification, which basically means the use of one method aiming at antigen detection or nucleic acid detection or cultivation of the agent. Confirmed identification requires at least two methods and unambiguous identification requires all three procedures and, if possible, an additional animal experiment to prove the pathogenicity of the agent. The latter level is, of course, reserved for stationary labs.

To ensure the quality of the diagnostic procedures and algorithms but also the evidential value of the results, accreditation of the stationary laboratory according to one of the pertinent standards, e.g. the European standard 15189 for medical labs, is mandatory as is the implementation of a quality management system in the mobile laboratory facilities.

13.2.3.5 Trace-Back Analysis and Attribution Investigation

Bioforensic attribution investigation and trace-back analysis of the source of the agent relies predominantly on the genetic analysis of the outbreak strain. In principle, it's taking a genetic fingerprint from the outbreak strain and comparing it with stored fingerprints of other strains of the same species by means of bioinformatic databases, into which the data of the individual strains have been integrated. The fingerprints can be linked to other available information on the individual strains like geographic regions where they typically occur, the strain history or known genetic characteristics of the respective species, e.g. clonality or mutation rates.

Various molecular typing methods can be used for this purpose [11], the most recent of which and the one with the highest discriminative power is whole-genome sequencing, which has become affordable and is a routine method now for outbreak analysis [12] in specialized lab units as BwIM's Microbial Genomics and Bioinformatics Group.

Bioinformatics is a discipline that contributes crucial and indispensable techniques to microbial forensics, e.g. algorithms for large-scale genome alignment and comparison or for the determination of gene composition, protein structure and the functional organization of genomes. It also provides methods to analyze the mechanisms and markers of pathogenicity and antibiotic resistance. The principal goal of bioinformatics in microbial forensics and diagnostics is to identify and characterize microorganisms at various levels of resolution (genus, species, clone, strain, substrain).

Microbial forensics does not only have the capacity to elucidate the origin of an outbreak strain, but also to reveal possible genetic manipulations that may be intended to enhance virulence or to introduce unexpected antibiotic resistance properties [13].

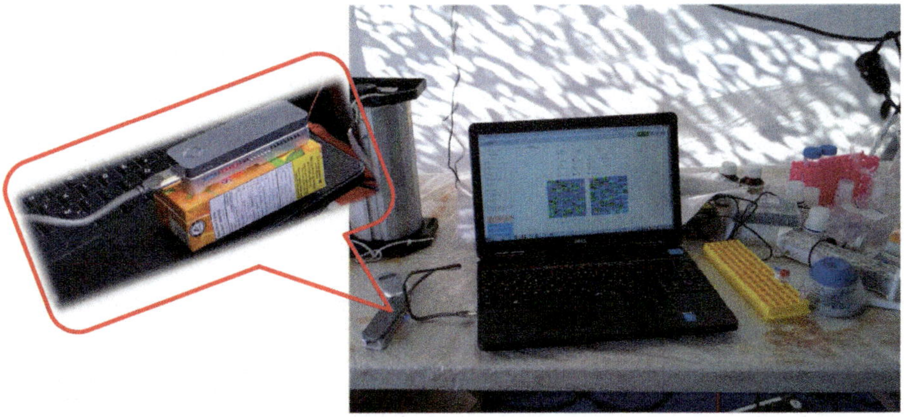

Fig. 13.3 The MinION™ sequencing device in the field laboratory during a NATO exercise

The latest development in microbial forensics is the use of portable third-generation sequencing devices like the MinION™ in the field [14], which allows sequencing independently from stationary laboratories and in close proximity to the epicenter of the outbreak (Fig. 13.3).

The value of such a trace-back analysis can be exemplified by an outbreak of anthrax in drug addicts caused by spore-contaminated heroine that had started in Scotland in 2010 and, after hitting several other countries, erupted in Bavaria in 2012. The method employed to investigate the outbreak strains was SNP analysis, which allows the attribution of strains to branches, clusters or groups by the analysis of single marker nucleotide positions in the genome that have changed during the course of evolution and have become characteristic for each clade. The outbreak strain could be attributed to a defined subcluster of the so-called A-strains. The closest relatives of the outbreak strain could be localized in Eastern Turkey close to the Afghan border, which is in concordance with the known route of heroine from its production site in Afghanistan via Turkey to Europe. Most probably the site where the contamination took place is somewhere in Eastern Turkey or in Afghanistan itself [15].

For the sake of completeness it should be mentioned here that besides the genetic signatures there are also chemical and physical signatures that may be relevant in microbial forensics [16]. For example, it is a proven fact that the chemical makeup of microorganisms (e.g. spores) reflects the environment (e.g. the medium) in which they have grown. Differences in the composition of media may, therefore, influence the protein profiles of the spores of Bacillus strains cultivated in them. Chemical substances added to media or spore preparations to ease aerosolubility may also provide forensic signatures. Methods to reveal such signatures include mass spectrometry and electron microscopy to name but a few. We are not going deeper into these aspects here because it is beyond the scope of medical bio-reconnaissance, which deals with patient specimens rather than with environmental samples. Microbial isolates cultivated from patients or animals will probably have lost the chemical and elemental signatures they may have originally had in the bioweapon. One of the major future challenges in the field of microbial forensics will be the implementation of a quality management system in order to ensure the legal usability of the data.

13.3 Conclusion

After the fall of the Berlin Wall in 1989 and entry into the "War on Terror" in 2001, the biothreat situation has fundamentally changed. A large-scale attack with airplanes spraying clouds of biological agents or throwing toxin-filled bombshells, the classical Cold War scenario, is no longer a major threat in a world where state-funded bioweapon programmes have largely been given up, although some states may still pursue hidden bioweapon programmes. Current threat analyses by intelligence services and security organisations instead point to the use of bacteria, viruses or toxins by personally (*biocrime*) or politico-religiously (*bioterrorism*) motivated groups or individuals. Modern biological defence, therefore, focuses on the development of concepts and strategies on how to minimize the damage caused by such scenarios to the population and/or to military forces. Early recognition of an attack is crucial as is the implementation of forensic procedures during the whole process of bio-reconnaissance. To prove a biological attack in a way that the investigative evidence is usable by the International Criminal Court or would justify a political or even a military reaction is, however, challenging. And, surprisingly, the paradigms have changed, as bio-reconnaissance is no longer a matter of the CBRN defence forces alone but requires sophisticated facilities of the Medical Services. Hence, this chapter outlines the rationale for the Medical Services of our NATO Armed Forces to enhance efforts in the field of bioforensics.

References

1. Read TD, Salzberg SL, Pop M, Shumway M, Umayam L, Jiang L, Holtzapple E, Busch JD, Smith KL, Schupp JM, Solomon D, Keim P, Fraser CM (2002) Comparative genome sequencing for discovery of novel polymorphisms in Bacillus anthracis. Science 296:2028–2033
2. Van Ert MN, Easterday WR, Simonson TS, U'Ren JM, Pearson T, Kenefic LJ, Busch JD, Huynh LY, Dukerich M, Trim CB, Beaudry J, Welty-Bernard A, Read T, Fraser CM, Ravel J, Keim P (2007) Strain-specific single-nucleotide polymorphism assays for the Bacillus anthracis Ames strain. J Clin Microbiol 45:47–53
3. Budowle B, Schutzer SE, Roger G, Breeze RG, Keim P, Stephen A, Morse SA (eds) (2010) Microbial forensics. Elsevier, Amsterdam
4. NATO DTP Lexicon (2017) https://nso.nato.int/nso/. Accessed 23 Sept 2017
5. Grundmann O (2014) The current state of bioterrorist attack surveillance and preparedness in the US. Risk Manag Healthc Policy 7:177–187
6. Holtherm HU (2012) Development of a multinational Deployment Health Surveillance Capability (DHSC) for NATO. Wehrmedizinische Monatsschrift 11–12

7. Grunow R, Finke EJ (2002) A procedure for differentiating between the intentional release of biological warfare agents and natural outbreaks of disease: its use in analyzing the tularemia outbreak in Kosovo in 1999 and 2000. Clin Microbiol Infect 8:510–521
8. Dembek ZF, Kortepeter MG, Pavlin JA (2007) Discernment between deliberate and natural infectious disease outbreaks. Epidemiol Infect 135:353–371
9. Wölfel R, Stoecker K, Fleischmann E, Gramsamer B, Wagner M, Molkenthin P, Di Caro A, Günther S, Ibrahim S, Genzel GH, Ozin-Hofsäss AJ, Formenty P, Zöller L (2015) Mobile diagnostics in outbreak response, not only for Ebola: a blueprint for a modular and robust field laboratory. Euro Surveill 20(44)
10. AEP-66 NATO Standard (2012) NATO handbook for Sampling and Identification of Biological, Chemical and Radiological Agents (SIBCRA) Edition 1, Version 1
11. Budowle B, Johnson MD, Fraser CM, Leighton TJ, Murch RS, Chakraborty R (2005) Genetic analysis and attribution of microbial forensics evidence. Crit Rev Microbiol 31(4):233–254
12. Gilchrist CA, Turner SD, Riley MF, Petri WA Jr, Hewlett EL (2015) Whole-genome sequencing in outbreak analysis. Clin Microbiol Rev 28(3):541–563
13. Antwerpen MH, Sahl JW, Birdsell D et al (2017) Unexpected relations of historical Anthrax strain. MBio. 2017 Apr 25;8(2). pii: e00440-17
14. Walter MC, Zwirglmaier K, Vette P, Holowachuk SA, Stoecker K, Genzel GH, Antwerpen MH (2017) MinION as part of a biomedical rapidly deployable laboratory. J Biotechnol 250:16–22
15. Hanczaruk M, Reischl U, Holzmann T, Frangoulidis D, Wagner DM, Keim PS, Antwerpen MH, Meyer H, Grass G (2014) Injectional anthrax in heroin users, Europe, 2000–2012. Emerg Infect Dis 2:322–323
16. Cliff JB, Kreuzer HW, Ehrhardt CJ, Wunschel DS (eds) (2012) Chemical and physical signatures for microbial forensics. Springer Book, Infectious Disease Series. Humana Press, New York

Chapter 14
The Role of Informal Digital Surveillance Systems Before, During and After Infectious Disease Outbreaks: A Critical Analysis

Avi Magid, Anat Gesser-Edelsburg, and Manfred S. Green

Abstract

Background One of the main limitations of traditional surveillance systems for disease detection is their inability to detect epidemics in real-time. In addition to syndromic surveillance, a number of informal digital resources have been developed. These systems are based on data collected through media sources such as news reports on the Internet, mailing lists, and RSS (Really Simple Syndication) feeds. The role of such systems at all stages of the epidemic remains unclear.

Methods A literature review was carried out on informal digital resources for infectious disease surveillance. We examined the source of information, the manner in which they process and disseminate the information, their role in each phase of disease outbreaks, and whether and to what extent these systems are capable of early detection and management of infectious disease epidemics.

Results Informal digital resources use similar sources of data for surveillance. However, they use different algorithms to create their output, and cover different geographic areas. In this regard, they complement each other with respect to information completeness. There is evidence in the literature on the systems' usefulness in communicating information to public health professionals, as well as to the general public during and after previous epidemics. Retrospective studies of some systems have shown a theoretical decrease in the time of epidemic detection compared to conventional surveillance. However, there is no evidence of the ability for real-time detection.

Conclusions Currently, there is little prospective evidence that existing informal systems are capable of real-time early detection of disease outbreaks. Most systems accumulate large amounts of information on a wide variety of diseases, making it

A. Magid (✉) · A. Gesser-Edelsburg · M. S. Green
School of Public Health, University of Haifa, Haifa, Israel

V. Radosavljevic et al. (eds.), *Defence Against Bioterrorism*, NATO Science for Peace and Security Series A: Chemistry and Biology,
https://doi.org/10.1007/978-94-024-1263-5_14

difficult to extract critical information. Presenting critical information clearly and precisely remains a challenge.

Keywords Infectious disease · Outbreak · Digital systems · Formal · Informal

14.1 Introduction

During the last few decades, emerging infectious diseases have become an increasingly important global public health problem and a major cause of morbidity and mortality. Emerging infectious diseases are characterized by a rapid increase in incidence or geographical range [17]. Examples of such outbreaks are the Severe Acute Respiratory Syndrome (SARS) originated from Asia in 2003, the Avian influenza H5N1, and the H1N1 2009 pandemic [17]. Effective surveillance systems for early warning of the outbreak are crucial. Traditional surveillance systems involve laboratory identification of the pathogen responsible for the disease. Such surveillance systems are passive in their nature, since they require a bottom-up process of identifying a possible infectious disease by clinicians, reporting to the appropriate authorities, confirming the disease by laboratory tests, and disseminating the information by the authorities.

Another prominent limitation of traditional surveillance systems is that they are not capable of detecting epidemics in real-time. Due to this limitation, new surveillance systems were developed. An impetus to developing these new surveillance systems was the adoption of the revised World Health Organization (WHO) International Health Regulations (IHR) of 2005, which required national capability for surveillance and reporting of both familiar and previously unfamiliar infectious diseases [25]. As a result, digital surveillance, or, digital systems for the detection of infection disease epidemics have been evolving dramatically. Morse has defined "digital disease detection" as "the use of the internet and computer technologies for collecting and processing health information, including outbreak reports and surveillance data" [17].

Digital systems can be classified into two types: formal and informal. These systems are based on syndromes rather than laboratory identification, without laboratory confirmation [17]. Formal digital systems are based on data arriving from formal organizations such as hospitals, healthcare clinics, and health agencies, whereas informal digital systems are based on data collected through media sources such as news reports on the Internet, mailing lists, and RSS (Really Simple Syndication) feeds, as well as data collected through official sources. Informal digital systems are characterized by their ability of mining, categorizing, filtering, and visualizing online information regarding epidemics [4]. They exploit the ease of using online information, as well as the freely available mapping technology to produce globally available information on ongoing infectious diseases, which may not be captured through traditional surveillance, and may be useful to governments and health agencies [4]. These systems are designed to function during all phases of disease outbreak, and are planned to increase sensitivity and timeliness. However, the role of such systems before, during and after infectious disease epidemics and, in particular, whether such systems are currently capable of early detection of epidemics remains unclear.

14.2 Methods

A literature review was carried out to compare informal digital systems with regards to their source of information, the manner in which they process and disseminate the information, their role in each phase of an epidemic, and whether and to what extent these systems are capable of early detection of epidemics. The systems evaluated were ProMED-mail, Global Public Health Intelligence Network (GPHIN), HealthMap, MediSys, EpiSPIDER, BioCaster, H5N1 Google Earth mashup, Avian Influenza Daily Digest and Blog, Google flu trends and Argus.

14.3 Results

14.3.1 Description of the Systems

14.3.1.1 ProMED-mail

ProMED-mail is "an internet based reporting system aimed at rapidly disseminating information on infectious disease outbreaks and acute exposures to toxins that affect human health, including those in animals and in plants grown for food or animal feed" [21] (ProMED-mail website). ProMED-mail receives information from a number of sources, such as media reports, official reports, online summaries and local observers. The reports are reviewed and investigated by ProMED-mail expert team, and then distributed by e-mail to ProMED subscribers, and published in ProMED-mail website (ProMED-mail website). In addition to filtering the received information, ProMED-mail expert team may also add related information from media, government and other sources [23]. ProMED-mail was proven as an efficient system during the 2003 outbreak of SARS, where information about points of outbreak, including additional information from a British Medical Journal article, was efficiently disseminated [23]. It should be stressed that ProMED-mail collects, filters, disseminates and archives it. They do not carry out formal analysis of the information although they provide some evaluation.

14.3.1.2 Global Public Health Intelligence Network (GPHIN)

The Global Public Health Intelligence Network (GPHIN) is a biosurveillance system developed by Health Canada in collaboration with the WHO. GPHIN receives as input, information about disease outbreaks arriving from news service items, ProMED-mail, electronic discussion groups and selected websites, and disseminates information to subscribers using the following decision algorithm. A relevance score is computed for each information item. Two thresholds are determined, high and low. If the item relevance score is greater than the high threshold, then it is immediately disseminated to subscribers. If the item relevance score is lower than the low threshold, then it is automatically "trashed". Otherwise (if the item relevance score is between the high and the low thresholds), the item goes through human analysis and then disseminated to subscribers [23].

A prominent limitation of GPHIN efficiency is its reliance on the time in which information about an outbreak or other event if published in one of GPHIN data sources. Nevertheless, GPHIN is considered efficient in providing earlier warning of events of interest to the international community compared with other systems, as 56% of the 578 outbreaks verified by WHO between July 1998 and August 2001 were initially picked up by GPHIN [23].

14.3.1.3 HealthMap

HealthMap is a freely accessible automated electronic information system aimed at facilitating knowledge management and early detection of infectious disease outbreaks by aggregating, extracting, categorizing, filtering and integrating reports on new and ongoing infectious disease outbreaks. Data on outbreaks are organized according to geography, time, and infectious disease agent [5].

HealthMap receives as input reports received from variety of electronic sources, including online news sources aggregated in websites such as Google News, reporting systems such as ProMED-mail, and validated official reports received from organizations such as the WHO [5, 11]. An internet search is performed by HealthMap every hour, 24 h a day, in order to obtain the required information. Search criteria include disease name (scientific and common), symptoms, keywords, and phrases. After collecting the reports, HealthMap uses text mining algorithms in order to characterize the reports. Characterization includes the following stages: (1) Categorization: reports are categorized according to disease and location and relevance is determined. (2) Clustering: similar reports are grouped together and exact duplicates are removed. Clustering is performed based on similarity of the report's headline, body text, and disease and location categories. (3) Filtering: reports are reviewed and corrected by an analyst, and then filtered into five categories – breaking news, warning, old news, context, and not disease related.

In order to reduce information overload and to focus on disseminating information regarding outbreaks of high impact, only reports classified as breaking news are overlaid on an interactive geographic map located on HealthMap site [5].

Among the users of HealthMap are the WHO, the US Centers for Disease Control and Prevention, and the European Center for Disease Prevention and Control, which use its information for surveillance activities [5, 11].

14.3.1.4 MedISys (Medical Intelligence System)

Medical Information System (MedISys) is an informal automatic public health surveillance system. MedISys is designed and operated by the Joint Research Center (JRC) of the European Commission, in cooperation with the Health Threat Unit at the European Union Directorate General for Health and Consumer Affairs and the University of Helsinki. MedISys collects its information from open-source news media, mainly articles from news pages. MedISys categorizes the collected information according to predefined categories and disseminates it to subscribed users by e-mail. The system also provides its user with features and statistics available on its website, including a world map in which event locations are highlighted, aggregated news count per each geographic location presented on graphs, and the most significant event location for the last 25 h. MedISys is available in 26 languages (the system collects information in 45 languages, but the website is available in 26 languages). Users can filter the information according to language, disease and location, as well as by outbreaks, treatments and legislations. MedISys users can also select articles into predefined categories, add comments to these articles, add information, and disseminate them to user-defined groups [12].

14.3.1.5 Argus

Argus is an informal biosurveillance system aimed at detecting and monitoring biological events that may be a global health threat to human, plant and animals. The system is hosted at the Georgetown University Medical Center (Washington, DC, United States), and funded by the United States Government. Argus collects information in 40 native languages from media sources, including printed newspapers, electronic media, Internet-based newsletters and blogs, as well as from official sources (the World Health Organization (WHO) and the World Organization For Animal Health (OIE). The system uses Bayesian analysis tools for selecting and filtering the collected articles. The process is performed by about 40 regional professional analysts, who monitor several thousand internet sources on a daily basis. By using Bayesian analysis tools, the analysts select reports from a dynamic database of media reports. Relevance is determined according to a specific set of terms and keywords applicable to infectious diseases surveillance. After filtering the information, events that may indicate the initiation of an outbreak are disseminated to the system users. Also disseminated are events that may require investigation [12, 20].

14.3.1.6 BioCaster

BioCaster is an informal surveillance system aimed to collect information on disease outbreaks, filter the information, and disseminate it to users. The system is a part of a research project developed and managed by the National Institute of Informatics in Japan, which involves five institutes in three countries. BioCaster focuses mainly on the Asia-Pacific region. The system collects information by using Really System Syndication (RSS) feeds from more than 1700 sources. Information is collected mainly from Google News, Yahoo! News, and European Media Monitor, filtered and disseminated in a fully automated manner with no human analysis in any stage. Filtered information (about 90 articles per day) is published in three languages (English, Japanese and Vietnamese). Articles are processed and disseminated every hour. In addition, BioCaster creates an ontology which covers approximately 117 infectious diseases and six syndromes. The ontology is produced in eight languages (English, Japanese, Vietnamese, Chinese, Thai, Korean, Spanish and French), and is used as an input to Global Health Monitor web portal, which offers its users maps and graphs of health-concerning events [12].

14.3.1.7 EpiSPIDER

The Semantic Processing and Integration of Distributed Electronic Resources for Epidemiology (EpiSPIDER) is a web-based tool which integrates information gathered from electronic media resources containing health information, as well as from informal surveillance systems, such as ProMED-mail. The aim is to enhance the surveillance of infectious disease outbreaks.EpiSPIDER uses ProMED-mail reports as an input, as well as health news sources that provide RSS feeds. By using natural language processing, it extracts location information from the input sources, and geocode them using the Yahoo and Google geocoding services. After a filtering process, the system generates summaries of ProMED reports (on a daily basis). These reports are available in the EpiSPIDER website [13].

14.3.1.8 H5N1 Google Earth Mashup

Google earth combines satellite images, aerial photography and map data to create a 3D interactive template of the world. This template can be used by anyone to add and share information about any subject that involves geographical elements. Nature (international weekly journal of science) uses Google earth to track the spread of the H5N1 avian flu virus around the globe, and to present a geographic visualization of the spread of H5N1 [19] (Nature website).

14.3.1.9 Avian Influenza Daily Digest

Avian Influenza Daily Digest is a digest produced by the United States government. The digest collects raw open source content regarding Avian influenza and disseminates it to subscribers. Material is disseminated without any processing. Users are encouraged to provide with updates and/or clarifications that will be posted in subsequent issues of the digest [2].

14.3.1.10 Google Flu Trend

Google Flu Trend is designed by Google Internet Company to be a near real-time tool for detection of influenza outbreaks. Google Flu Trend exploits the fact that millions of people worldwide search online for health-related information on a daily basis. The tool was designed based on the assumption that there is an association between the number of people searching for influenza-related topics and the number of people who actually have influenza symptoms, and therefore, an unusual increase in the number of people searching for influenza-related topic on the web may simulate an increase in influenza syndromes. Studies performed by Google and Yahoo have shown that plotting data on searches using influenza-related keywords has led to an epidemic curve that closely matched the epidemic curve generated by traditional surveillance of influenza [4]. Google Flu Trends analyzes a fraction of the total Google searches over a period of time, and extrapolates the data to estimate the search volume. The information is displayed in a graph called "search volume index graph". It is claimed by the tool's designers that, according to tool testing, it can detect outbreaks of influenza 7–10 days before it is detected by conventional CDC surveillance [4].

14.3.2 Comparison Between Systems

All the studied digital resources use similar sources of data – official reports, as well as media reports, including global media resources, news aggregators, eyewitness reports, internet-based newsletters and blogs. However, they use different algorithms to create their output, and cover different geographic areas. In addition, existing digital resources are different in the manner they filter and analyze the information and may create different output. Therefore they complement each other with respect to information completeness.

14.3.3 The Role of Informal Digital Systems in Each Phase of the Epidemic

14.3.3.1 Before the Epidemic (Early Detection)

Retrospective studies of some systems have shown a theoretical decrease in the time of outbreak detection compared to conventional surveillance. However, evidence of such ability in real time is sparse and unclear. Chan et al. [8] have analyzed the average interval between the estimated start of the outbreak to the earliest date of discovery and publication, using WHO confirmed outbreak reports, as well as ProMED-mail, GPHIN and Healthmap reports. Analysis showed a decrease in intervals over 14 years, which was partially attributed to the emergence of informal digital resources [8]. A retrospective study of Argus reports on respiratory disease in Mexico showed a significant increase in reporting frequency during the 2008–2009 influenza season relative to that of 2007–2008. The authors suggest that, according to these retrospective results, respiratory disease was prevalent in Mexico and reported as unusual much earlier than when the H1N1 pandemic virus was formally identified. However, its connection with the 2009 pandemic is unclear [20].

The Google Flu Trends tool was also retrospectively tested. According to retrospective testing, influenza epidemics can be detected by using Google flu trends tool 7–10 days before it is detected by conventional surveillance [6, 7], however, there are still no prospective evidence to such capability. A retrospective study from China reported that Google flu trend search data are correlated with traditional methods of surveillance [14]. Another retrospective study tested the real-time detection ability of six informal digital systems, including Argus, BioCaster, GPHIN, HealthMap, MedISys and ProMED-mail. Data from these systems were used to detect epidemics and compared to official data. Results suggested that all tested systems have shown retrospective real-time detection ability. Moreover, it was found that the combined expertise amongst systems provided a better early detection [3].

Unlike retrospective evidence, prospective evidence of informal digital systems capability for early detection of epidemics is sparse. Some epidemics have been claimed to be first reported by ProMED-mail, before they were officially reported by the WHO [15]. These reports were proved to be reliable, since they were later confirmed by the WHO. However most of the reports were first published by ProMED-mail not because the information was not available to the WHO by this time, but because the WHO was not authorized to publish them due to lack of conformation [15]. The SARS in China (February 10, 2003) is the best known outbreak first reported on ProMED-mail [17].

A detection in real-time was also demonstrated by GPHIN during the SARS outbreak of 2002. GPHIN detected SARS and issues the first alert to the WHO more than 2 months before it was first published by the WHO [16, 18, 24]. However, the time between the GPHIN alert and the first time it was reported by the WHO implies that the whole detection process was not shortened due to the GPHIN alert.

Retrospective reviews of the polio outbreak of 2013 and 2014 and the Ebola outbreak of 2014 showed that informal digital detection preceded official detection by an average of 14.6 days. For example, ProMED and GPHIN reported the polio epidemic in Cameroon in 2013 23 days after the outbreak began, where the official WHO report was published 51 days after the outbreak began [1]. However, the digital systems detection did not contribute in real-time to the whole process of outbreak detection and declaration. Hence, in real-time it is not an early detection.

14.3.3.2 During the Outbreak

There is evidence in the literature on the systems' usefulness in communicating the information during previous outbreaks to public health professionals, as well as to the general public. ProMED-mail and GPHIN had critical roles in updating public health officials about the SARS outbreak in 2002 [4]. Such systems are also capable of providing officials, clinicians and the general public with guidance to medical decision making, including the importance of vaccination and other preventive actions [4]. The first report on SARS on February 10, 2003 published by ProMED-mail, and the hundreds of subsequent ProMED-mail reports have helped health professionals worldwide to gather critical details regarding SARS, and by this to recognize SARS and discover its cause [15]. Assessment of correlation between Healthmap reports and official government reports reported during the first 100 day of the 2010 Haitian Cholera outbreak has confirmed that data yielded from informal digital systems were well correlated with data officially reported from the Haitian health authorities. Moreover, this study has shown that informal digital systems are capable of being used at the early stages of an outbreak not only as an indicator of the outbreak occurrence, but also as a predictive tool by providing a reliable estimation of the reproductive number, a major epidemic parameter [9].

14.3.3.3 After the Outbreak

There is no evidence in the literature of the use of informal digital systems after an epiodemic. Nevertheless, we believe that data collected during outbreaks through informal digital systems are being used by public health agencies for retrospectively studying the dynamics of epidemic, and for drawing conclusions about the management of the epidemic.

14.4 Discussion

There has been impressive progress in the development of informal digital systems for disease surveillance. Informal digital systems are widely used by the general public, as well as by health officials. A good example is the GORAN digital system (the Global Outbreak and Response Network) developed by the WHO, which gather information from number of sources both governmental and informal, including GPHIN and ProMED-mail [17].

One of the most prominent suggested advantages of the digital systems is their functioning in early notification of infectious disease outbreaks, before the official notifications, and their contribution to the epidemiological investigation of the disease before official data are available. During epidemics, data gathered and disseminated through official public health authorities are usually not available to public health officials and to policy makers for some time, sometimes due to political and logistic limitations. This period of time is critical for estimating the epidemic dynamics and implementing the response plan [9]. Unlike official data, data collected by digital systems are available in near real-time, and may be used for epidemiological assessment.

A mandatory requisite for the use of digital systems data for epidemiological investigation of an outbreak is the reliability of the data, as well as their equivalence to official data. In other words, there should be a match between the number of cases derived from the informal data and the number of cases officially reported by public health authorities. Indeed, our results have pointed out an example in which a correlation between digital systems data and official data in the first stages of epidemic was confirmed in the data collected from Healthmap regarding the 2010 Haitian Cholera. However, as mentioned by the authors, epidemiological measurements using digital systems data should be also tested in other epidemics, in order to confirm the method's reliability [9]. The fact that the number of subscribers to digital systems is increasing each year [15] makes these systems an efficient tool for globally spreading the information, as well as a tool for epidemiological investigation, complementary to official data. However, despite their theoretical advantage over traditional surveillance, there is no evidence in the literature that information collected through digital system had affected public health policy makers.

Although we did not find evidence in the literature, we believe that digital systems may also contribute to the public health community after the outbreak ends. The abundance of reports collected and disseminated by these systems during outbreaks creates an epidemiological reservoir, which, due to its availability worldwide, may be used for a post-pandemic investigation and conclusion making.

As for early detection of infectious disease outbreaks, we did not find any prospective evidence showing the capability of digital systems of detection infectious disease outbreaks in real-time. Our results are consistent with some other studies conclusions, pointing out that currently digital systems are not capable of detecting an outbreak [17, 4]. Although there is evidence of informal digital systems publishing reports on outbreaks before official detection (such as in the Polio outbreak of 2013 and 2014 pointed out in the results section) [1], these reports did not actually affect the process of detection. The formal process of detection includes receiving the information, processing the information and using the information. The early digital systems reports were not used in any of the detection phases and did not change the process. It may be viewed as an analogue to screening tests which are effective only if they are capable of changing the natural history of a disease. Since there is no evidence of informal digital systems capable of changing the "natural history of outbreak" so far, they cannot be considered useful for early detection.

Informal digital systems may also have an important role in disease surveillance. Incorporating informal digital systems into existing formal systems may improve their performance. A study in the United States showed that combining information gathered from informal digital systems with information received from the Texas Influenza-Like-Illness Surveillance Network (ILINet) improved the ability of predicting hospitalizations due to influenza [22]. Another study in the United States showed a good correlation between Google flu searches and emergency department influenza-like illness visits [10].

Moreover, since digital sources usually contains data not captured through traditional methods, they are used by public health organizations, including the Global Outbreak Alert and Response Network of the WHO, which uses digital sources for surveillance on a daily basis [4].

However, the usage of digital systems as a surveillance tool may have some limitations. First, most systems accumulate a huge mass of information on a large variety of diseases, making it difficult to extract critical information. In other words, no integration of the information is performed to yield useful information. The challenge is to present critical information clearly and concisely. Second, digital systems are less specific than traditional surveillance systems, mostly due to false alarms, misinformation and information based on rumors [9, 17]. Therefore, they may not be solely used but as a complementary tool for traditional surveillance systems [17]. A third limitation is the lack of a response system to early warnings. With the lack of such a system, early warning is not useful, as no practical action is followed by the publication of the information. Such a response system may include triggers and decision criteria, which would lead to an appropriate and proportionate response to the threat [17].

To summarize, considerable efforts and resources have been invested in the development of informal digital system for detection of infectious disease outbreaks. As a result, a new generation of informal digital systems has emerged. The most prominent advantage of such systems is their ability to report on an outbreak in near real-time, or, in other words, before the information is officially reported, and by this to be used by public health decision makers for epidemiological assessment and preparation for the pandemic. Currently there is no evidence in the literature for their capability to detect an outbreak at its onset. In addition, there are no hard data to prove the benefits of using such systems before and during an outbreak. We do not believe that they can be used to identify early cases, but should be used as a support system for describing the spread of the disease. The challenge is to empirically assess the efficiency of informal digital systems and their use for decision making and interventions during crisis, as well as to test the systems' sensitivity and specificity. A more general informal system, which provides syndromic-based analysis of reports disseminated by all currently existing systems, may be the next step toward disease outbreak detection based on informal systems.

References

1. Anema A, Kluberg S, Wilson K, Hogg RS, Khan K, Hay SI et al (2014) Digital surveillance for enhanced detection and response to outbreaks. Lancet Infect Dis 14(11):1035
2. Avian Influenza Daily Digest website: http://aidailydigest.blogspot.co.il/2008/10/dni-avian-influenza-daily-digest_09.html
3. Barboza P, Vaillant L, Le Strat Y, Hartley DM, Nelson NP, Mawudeku A, Madoff LC, Linge JP, Collier N, Brownstein JS, Astagneau P (2014) Factors influencing performance of internet-based biosurveillance systems used in epidemic intelligence for early detection of infectious diseases outbreaks. PLoS One 9(3):e90536
4. Brownstein JS, Freifeld CC, Madoff LC (2009) Digital disease detection – harnessing the web for public health surveillance. N Engl J Med 360:2153–2157
5. Brownstein JS, Freifeld CC, Reis BY, Mandl RK (2008) Surveillance Sans Frontieres: internet-based emerging infectious disease intelligence and the HealthMap project. PLoS Med 5:e151
6. Carneiro HA, Mylonakis E (2009) Google trends: a web-based tool for real-time surveillance of disease outbreaks. Clin Infect Dis 49(10):1557–1564
7. Carneiro HA, Mylonakis E (2013) Google trends: a web-based tool for real-time surveillance of disease outbreaks. Surfing Web 2009(49):1557
8. Chan EH, Brewer TF, Madoff LC, Pollack MP, Sonricker AL, Keller M, Freifeld CC, Blench M, Mawudeku A, Brownstein JS (2010) Global capacity for emerging infectious disease detection. Proc Natl Acad Sci 107(50):21701–21706
9. Chunara R, Andrews JR, Brownstein JS (2012) Social and news media anables estimation of epidemiological patterns early in the 2010 Haitian Cholera outbreak. Am J Trop Med Hyg 86(1):39–45
10. Dugas AF, Hsieh YH, Levin SR, Pines JM, Mareiniss DP, Mohareb A, Gaydos CA, Perl TM, Rothman RE (2012) Google Flu trends: correlation with emergency department influenza rates and crowding metrics. Clin Infect Dis 54(4):463–464
11. Freifeld CC, Mandl KD, Reis BY, Brownstein JS (2008) HealthMap: global infectious disease monitoring through automated classification and visualization of internet media reports. J Am Med Inform Assoc 15:150–157
12. Hartley DM, Nelson NP, Walters R, Arthur R, Yangarber R, Madoff L, Linge JP, Mawudeku A, Collier N, Brownstein JS, Thinus G, Lightfoot N (2010) Landscape of international event-based biosurveillance. Emerg Health Threats J 3:e3
13. Herman Tolentino M, Raoul Kamadjeu MD, Michael Matters PhD MPH, Marjorie Pollack MD, Larry Madoff MD (2007) Scanning the emerging infectious diseases horizon-visualizing ProMED emails using EpiSPIDER. Adv Dis Surveil 2:169
14. Kang M, Zhong H, He J, Rutheford S, Yang F (2013) Using google trends for influenza surveillance in south China. PLoS One 8(1):e55205
15. Madoff LC, Woodall JP (2005) The internet and the global monitoring of emerging diseases: lessons from the first 10 years of ProMED-mail. Arch Med Res 36:724–730
16. Milinovich GJ, Williams GM, Clements AC, Hu W (2014) Internet-based surveillance systems for monitoring emerging infectious diseases. Lancet Infect Dis 14(2):160–168
17. Morse SS (2012) Public health surveillance and infectious disease detection. Biosec Bioterror Biodef Strateg Pract Sci 10(1):6
18. Mykhalovskiy E, Weir L (2006) The global public health intelligence network and early warning outbreak detection: a Canadian contribution to global public health. Can J Public Health:42–44
19. Nature website: www.nature.com/avianflu/google-earth
20. Nelson NP, Brownstein JS, Hartley DM (2010) Event-based biosurveillance of respiratory disease in Mexico, 2007–2009: connection to the 2009 influenza A (H1N1) pandemic? EuroSurveillance 15(30)

21. ProMED-mail website: http://www.promedmail.org/
22. Scarpino SV, Dimitrov NB, Meyers LA (2012) Optimizing provider recruitment for influenza surveillance networks. PLoS 8(4)
23. Wagner MM, Moore AW, Aryel RM (2006) Book: handbook of biosurveillance 301
24. Wilson K, Brownstein JS (2009) Early detection of disease outbreaks using the Internet. Can Med Assoc J 180(8):829–831
25. World Health Organization (2008) International health regulations (2005), 2nd edn. World Health Organization, Geneva

Chapter 15
Strategic Aspects of Countering Bioterrorism

Katarina Strbac and Branislav Milosavljevic

Abstract Article is devoted to phenomena of bioterrorism which is not new threat to security, technique and methods of combating this phenomena should be considered carefully in societies today. Authors in article emphasis that first step in combating bioterrorism are strategies which parts clearly explain what should be done. Legal frameworks, common understanding of challenges and threats, standardized rules of operation, improved exchange of information, increased capability to prevent biological attacks are procedures as an integral parts of strategies for combat weapon for mass destruction including bio weapon.

Article is consist of: basic terms which remind us what exactly biological weapon is and explaining that malevolent application of biological agents in terrorist acts to cause infectious diseases of civilians or military personnel, animals and plants, also international legal framework concerning biological terrorism, reasons for strategic approach for countering bioterrorism, as an example how control and prevent this phenomenon. In addition, the strategy can be seen as an expression of the evolution of the control of biological weapons focusing on the projection of future manifestations of bioterrorism, in order to take optimal measures in countering this phenomenon. Different international initiative are good tool for developing strategies as a first step in understanding and preventing use of biological weapon. In article are explained several regional initiatives and their action regarding bio weapon including national approaches to this security problem. Without contemporary thinking and acting, this issue cannot be solved in future years, threat of biological weapon will grow if humanity doesn't take serious measures in prevention and combating this phenomena.

K. Strbac (✉)
Ministry of Defence of the Republic of Serbia, Belgrade, Serbia

B. Milosavljevic
Strategic Research Institute, Ministry of Defence of the Republic of Serbia, Belgrade, Serbia

V. Radosavljevic et al. (eds.), *Defence Against Bioterrorism*, NATO Science for Peace and Security Series A: Chemistry and Biology,
https://doi.org/10.1007/978-94-024-1263-5_15

15.1 Introduction

In the contemporary world the issue is not whether, but when a bioterrorist attack is going to occur. What used to be a theoretic possibility during the Cold War became reality during the last decade of the twentieth and the beginning of the twenty-first century, in the wake of the attacks on Tokyo subway system in Japan (1995) and the American "Anthrax crises" (2001). The concerns over non-state actors obtaining capability to intentionally release biological pathogens have increased considerably, although the use of conventional instruments of terrorism has not diminished. The possibility that terrorists might apply biological weapons in the predictable future represents a great concern of governments, international organizations and public worldwide, having in mind a clear risk of multiplication which would potentially increase the effects of such terror-motivated acts. With regard to the aforementioned, several international organizations and a number of countries have decided to develop a comprehensive strategic approach, with the aim of preventing this type of threat to national, regional and global security. Harmonized strategies should lead us to complementary legal frameworks, common understanding of challenges and threats, standardized rules of operation, improved exchange of information and finally, increased capability to prevent biological attacks or limit their impact on the targeted territory.

The fact is that the twentieth century is filled with threats which prominent before, most notably terror attack. Even though there is a highly developed consciousness concerning terror attacks, countries must invest a lot of time and effort for their security, especially in the prevention of terror attacks. In this paper, the authors wish to inform the reader about the need to create a strategic framework to prevent the weapon of mass destruction proliferation, especially concerning countries of South-eastern Europe which are facing multiplied threats to their safety. The Republic of Serbia, as a part of South-eastern Europe, joins the total effort against mass weapon proliferation, including actions against bioterrorism. Creation and adoption of adequate strategies on a national level are a good starting point for the improvement of fighting capacities on a national, regional and global level against the usage of all weapons of mass destruction.

15.2 Background

The constant progress of science, linked with the fact that biological weapons possess features which (under certain circumstances) make them suitable for violent political purposes, has created conducive environment for abuse of biotechnologies by terrorist organizations.

Human history records numerous cases of biological agents being misused by armies in warfare. Such ambitions are probably as old as the mankind itself,[1]

[1]Costa, H., & Baños, J. (2016). *Bioterrorism in the literature of the nineteenth century: The case of wells and the stolen bacillus. Cogent Arts & Humanities, 3*(1).

although practical capabilities of belligerents to defeat enemies using disease-producing materials used to vary in various epochs, in direct co-relation with available or lacking levels of knowledge. Notable examples of biological weapons application were described by ancient historians and chronologists, covering Persian, Greek and Roman campaigns. Hannibal in 190BC used clay pots filled with snake venom when targeting enemies.[2] Perhaps the most impressive example of biological weapons application was the intentional infecting of the harbor defenders of Crimea in 1346, when the Mongols catapulted plague-infected corpses over the fortress ramparts. Thus the disease, known in history as the "Black Death" had spread through Sicily, Corsica and Genoa, continued throughout Europe, eventually killing approximately 25 million people.

During World War I, the use of biological weapons was not recorded, as all the conflicting parties considered such attacks as worse than inhumane. Two decades later, in World War II, the Japanese used swarms of infected fleas, intending to infect Chinese soldiers with Bubonic plague. The exact figures cannot be determined, due to the fact that a number of the infected and sick people were already present in the war-torn area.

It is hard to predict a possible usage of biological weapon, as in the modern history of warfare there has not been enough evidence that they have been used intentionally. Therefore, numerous estimates have been based exclusively on natural outbreaks and experimental laboratory models. Development, production, storage and use of biological weapons are prohibited by Conventions and international law.[3] Despite all efforts, the threat of possible deliberate use of biological agents has increased since the end of the Cold War to the present.[4] During the half a century long confrontation of super powers, the primary global security concern was the nuclear war threat of. Bioterrorism was also perceived as a potential challenge, but with a limited impact on public imagination.[5] It is very hard to determine the exact number of the state actors who own bio-weapons, as both the possession and research on biological weapons can often be justified by a necessity to keep them for defensive purposes, which neither prevents the production of offensive biological agents, nor negates the need to establish protective measures. In other words, a shift from defensive to offensive biological weapons programs can be performed easily and quickly. On the other hand, the research of biological agents has contributed to gaining new scientific knowledge in microbiology, pathology, genetics and other fields.

The development of science in the concerned fields, among other results, has enabled scientists to permanently change structures of pathogenic microorganisms, increasing their infectious capability and resistance. Unlike the nuclear and chemi-

---

[2] Foster T George, Focus on Bioterrorism, Nova science Publishers Inc., 2006.

[3] International Convention for the Suppression of the terrorist bombings (1997).

[4] International Convention for the Suppression of the financing of terrorism (1999) United Nations Security Council Resolution 1540 (2004).

[5] Clarke, S. C. (2002). Bioterrorism: An overview. *British Journal of Biomedical Science, 59*(4), 232–4.

cal weapons, it is difficult to detect a biological weapon in the early stage. The nuclear weapons are complex, expensive and require advanced transmission systems. Chemical weapons are easier and cheaper to manufacture, but difficult to deliver to target areas. Biological weapons are fundamentally different from other weapons of mass destruction. While nuclear and chemical weapons have instant effects, biological agents require hours to several days or even weeks of incubation before they can cause death. Biological weapons are relatively cheap and easy to manufacture, which makes them attractive for terrorist purposes. Another aggravating circumstance is that bio-weapons may be secretly produced, making the timely detection of their presence very difficult, which creates an additional risk to potentially affected countries. Also, possible targets of a biological attack do not have to be humans. They can be aimed at domestic animals, food or agricultural production. Bio-weapons can cause unpredictable psychological consequences for the defending forces and the civilian population, such as mass panic and loss of morale. Inability to provide adequate protection to citizens may be followed by additional psychological effects. Emergence of panic is especially dangerous, which gives particular importance to the existence and adequate training of biodefense units.

15.3 Biological Weapons: Basic Terms

Biological weapons include microbes and other biological agents, or toxins (whatever their origin or method of obtaining might be), kept in possession which is not intended for prophylactic, protective or other peaceful purposes. The term may also be applied to weapons, equipment and other means or methods of dissemination of the aforementioned agents, used with hostile intent or during the war.

15.3.1 Definition

According to WHO definitions:

- biological agent is a "micro-organism (or a toxin derived from it) which causes disease in personnel, plants, or animals or causes the deterioration of materiel".
- biological warfare "is the use of biological agents to cause the loss of people and livestock, as well as damage to plants and materials";
- biological defense "includes the established methods, plans and procedures and implemented measures of defense against biological attacks".[6]

The terms biological weapons and biological warfare first appeared in official use after World War II, at the meeting of the UN General Assembly held in 1947,

[6] http://www.who.int/csr/delibepidemics/chapter3.pdf, Public health response to biological and chemical weapons—WHO guidance.

when biological, nuclear and chemical weapons were included in the group of the weapons of mass destruction. The concept of weapons of mass destruction came into use at the end of the Cold War in the United States as a common term replacing previous formulations of nuclear, biological and chemical weapons. It was considered to be a more adequate approach, as all of these weapons significantly differ in their effects and the principles of their use for military purposes. In addition, each of these categories was regulated by different rules in terms of arms control and proliferation. Introduction of a common concept came in direct correlation with the new tasks of the Armed Forces of the United States, defined after the end of the Cold War. This was also the period during which, at the same time, proliferation ban for all the three named categories of weapons became one of primary tasks in foreign and security policy of many countries.[7]

The notion of the weapons of mass destruction was derived from the UN Recommendation (Committee on Conventional Armaments 1948).The term was used for weapons with the common feature of causing large destructive effects and huge human casualties or mass starvation. Since then, Biological Weapons have been perceived as a potentially most dangerous means of mass destruction which may be applied on humans, animals or plants with unforeseeable consequences. As defined by the Convention on the Prohibition of the Development, Production and Stockpiling of Bacteriological (Biological) and Toxin Weapons and on their Destruction, commonly known as the Biological Weapons Convention (BWC) or Biological and Toxin Weapons Convention (BTWC), (1972–1975),[8] biological agents are classified as living organisms that are naturally derived, or artificially produced, which can cause illness or death of people, animals and plants, depending on the effects and the ability of reproduction in the human, animal or plant.

Pathogenic microorganisms are bacteria, viruses, fungi and protozoa, natural or modified by genetic engineering or other biotechnological process, as well as their poisons, if their purpose is not peaceful, causing an epidemic (on humans) epizootic (on animals) or epiphytotic (on plants). According to the NATO definition, biological and toxin warfare agents[9] are microorganisms and toxins derived from them with the purpose of causing disease in humans, animals and plants or degradation of environment.

The agents are derived from living micro-organisms or their products, and their incorporation into various types of weapons becomes a biological weapon. For a more complete understanding of the subject area, it is necessary to define the concept of proliferation, usually related to the weapons of mass destruction. The term

---

[7] Ostfield, M. L. (2004). Bioterrorism as a foreign policy issue. The SAIS Review of International Affairs, 24(1).

[8] https://www.un.org/disarmament/geneva/bwc/. The Biological Weapons Convention.

[9] Chevrier, Marie Isabelle; Chomiczewski, Krzysztof; Garrigue, Henri, eds. (2004). The Implementation of Legally Binding Measures to Strengthen the Biological and Toxin Weapons Convention: Proceedings of the NATO Advanced Study Institute, Held in Budapest, Hungary, 2001. Volume 150 of NATO science series: Mathematics, physics, and chemistry (illustrated ed.). Springer.

comes from the French word *proliferation*, which denotes germination, sprouting, budding and in everyday speech, the expansion or the spread of such weapons, indicating that it is not used only for countries that initially acquire the weapons of mass destruction but it also indicates a qualitative improvement of the existing arsenal of a country. In the past, the notion of "horizontal proliferation" was used to denote the former and "vertical proliferation", to denote the latter. The opposite process of proliferation, non-proliferation is used to indicate the renunciation of the existing state arsenal of the weapons of mass destruction.

15.3.2 Biological Weapon Characteristics

Particular danger of biological weapons lies in a number of biological agents that are already found in nature and are potential biological weapons, making it difficult to distinguish between the situations in which a disease is deliberately spread, and the situations that occur naturally.

The indisputable fact that there are plenty of viruses and pathogenic organisms found in nature does not mean that they are all suitable for terrorist purposes. In the history of mankind, biological weapons have often been used as weapons of war or for achieving other goals, although their use has always been considered shameful. Therefore, the question is why, despite the universal public condemnation, some actors do not give up, but keep, produce, improve and apply them in a given situation.

The answer could be that there are numerous characteristics that make biological weapons attractive to use, the most significant being:

(a) simple production, because certain biological agents are easy to produce in modestly equipped microbiological laboratories: all that is required for the reproduction of bacterial culture is a nutrient medium and an incubator thermostat;
(b) low-cost production, related to the aforementioned; According to some calculations from a few decades ago, the cost of achieving a particular effect ("neutralizing manpower") on the surface of 1 km^2 by using various types of weapons are: conventional – $2000, nuclear – $800 chemical – $600, biological only 1 dollar;
(c) bioterrorism or biological aggression is very difficult to prove if there is no convincing epidemiological evidence or material; With the knowledge of epidemiological and ecological characteristics of an area, professional users of biological weapons can cause illness on a smaller or larger scale that cannot be distinguished from naturally occurring epidemics;
(d) effective implementation, because a 1 kg of anthrax spores disseminated as an aerosol can cover an area of 100 km^2 and lead to the death of 50% of people;
(e) specific effects on humans, animals or plants, without causing significant material damage, destruction and without significant environmental consequences;

(f) causing mass morbidity – death; it depends, mainly, on the type of pathogen and the route of the dissemination of the biological agent; the most appropriate agents that can be disseminated by air (aerosol), and the ones with a possibility of subsequent inter-human transmission (small pox virus);
(g) causing panic, political instability, disruption of health and other services, as well as disruption of normal activities with all the resulting consequences;
(h) the problem of required fast detection and identification of the applied agents, in order to establish adequate measures to neutralize biological attack, and provide adequate treatment for the exposed patients, as well as pre-exposure *prophylaxis*.

Biological weapons can penetrate the body in three different ways: inhalation represents the most likely way; inhaling infectious organisms or toxins found in the air. Another way is through ingestion or swallowing, allowing the infection or intoxication of the digestive organs. Absorption through the mucous membrane exposure, through the skin or as a result of wounds or scratches is the third possibility. In addition, biological weapons can damage the material resources or render them unusable.

15.4 International Legal Framework

The first attempt to control the use of chemical and biological weapons in armed conflicts, perceived as a growing threat, incomparable to all previously known types of weaponry, came at the end of the nineteenth and the beginning of the twentieth century.

The Hague Conventions of 1899[10] and 1907 came as a result of several international treaties and declarations negotiated at two international peace conferences, held in the Hague (the Netherlands). The first Hague Conference was held on the initiative of the Russian emperor Nicholas II Romanov "with the object of seeking the most effective means of ensuring to all peoples the benefits of a real and lasting peace, and, above all, of limiting the progressive development of existing armaments."[11] From May 18th to July 29th, representatives of 26 governments were trying to reach an agreement on limitation or reduction of armaments. The attempt was not completely successful, although three conventions, several other acts and the Final Protocol[12] were adopted. The Second Hague conference lasted from June 15th until October 18th 1907. The authoritative statements within the Final Act were signed by the delegates, but not ratified by the participating states. For that

[10] International Committee of the Red Cross web database: https://ihl-databases.icrc.org/ihl/INTRO/150?OpenDocument

[11] https://ihl-databases.icrc.org/ihl/INTRO/145?OpenDocument

[12] The Final Protocol of the First Geneva Conference (1899): https://ihl.databases.icrc.org/applic/ihl/ihl.nsf/Article.xsp?action=openDocument&documentId=8FCF14D950797012C12563CD005 15C0A

reason, they never became binding. The third conference was planned for 1914/15, but was never held, due to the start of World War I.

Seven years after the end of the World War 1, there was another attempt. The Geneva Protocol (*Protocol for the Prohibition of the Use in War of Asphyxiating, Poisonous or other Gases, and of Bacteriological Methods of Warfare) was a treaty that prohibited* the *use of chemical and biological weapons in international armed conflicts.*[13] It was signed under the framework of the League of Nations[14] on June 17th 1925 and entered into force in 1928. This project, although well-intended, could not produce a long-lasting effect, due to the fact that the organization under which it was adopted and supposed to be implemented, gradually collapsed after 1933. Japan and Germany left the League of Nations first (1933), Italy followed their example in 1937 and only 2 years later, World War II started.

The third attempt came almost a half of century later, in the middle of the Cold War. The Convention on the Prohibition of the Development, Production and Stockpiling of Bacteriological (Biological) and Toxin Weapons and on their Destruction[15] is often quoted as "The Biological Weapons Convention (BWC)". It was the first multilateral disarmament treaty that banned the development, production and stockpiling of an entire category of the weapons of mass destruction. The document was signed on 10 April 1972 and entered into force in 1975. Followed by **six** review **conferences**, General Assembly Resolutions and statements of the Secretary General of the UN, the Convention remained the supreme global framework. By the end of 2016, it was signed by 173 states and ratified by 22.

15.5 Bioterrorism

The term bioterrorism has multiple meanings. It primarily covers malevolent application of biological agents in terrorist acts to cause infectious diseases of civilians or military personnel, animals and plants, being spread in the form of an epidemic or pandemic. Biological agents can be used to spread infection through the air, water or through food. From the perspective of a possible use of biological weapons for terrorist purposes, potential terrorist organizations have a full range of harmful agents available.

[13] Protocol for the Prohibition of the Use in War of Asphyxiating, Poisonous or Other Gases, and of Bacteriological Methods of Warfare, 1925 Geneva Protocol. https://www.un.org/disarmament/wmd/bio/1925-geneva-protocol

[14] Encyclopaedia Britannica, internet edition, League of Nations: https://www.britannica.com/topic/League-of-Nations (30/06/2017).

[15] United Nations Office for Disarmament Affairs (UNODA) official web site: https://www.un.org/disarmament/wmd/bio

15.5.1 Potential Agents of Bioterrorism in Contemporary World, According to Clarke (2002)[16]

Agent	Disease
Bacillus anthracis	Anthrax
Francisella tularensis	Tularaemia
Yersinia pestis	Plague
Variola virus	Smallpox
Hemorrhagic viruses	Viral hemorrhagic fever
Botulinum toxin	Botulism
Brucella spp.	Brucellosis
Vibrio cholerae	Cholera
Burkholderia pseudomallei	Glanders
Coxiella burneti	Q fever

In addition to that, a special "mitigating circumstance" is their availability, especially in clinical and microbiological laboratories and other scientific institutions. They have a short incubation period, are very contagious and consistently act in small doses. With a very low cost, accessible equipment and widely available knowledge, the production of these agents is very easy.

The threat of terrorism is different from those of the past in the changed tactics, increased destructiveness, the introduction of professionally planned and coordinated attacks, as well as the transnational character of the operations. There are growing discussions on the terms *postmodern terrorism* or *super terrorism*. With the aim to draw attention to the use of weapons of mass destruction for the purpose of terrorist attacks. Today, there is a much greater danger of their use by various organizations, cults and individuals. In this sense, the term "bio-terrorism", which is defined as the violent use of biological agents for political, religious, environmental or other ideological reasons, regardless of their moral or political justification. Some of these terms are different from the term "bio criminal act" and identify the use of biological agents for reasons not related to ideology.

Authors find it appropriate to recall two significant cases which included intentional biological threat to public safety and influenced later strategic reflection of global and national response to bio-terrorism. On March 20 1995, Aum Shinrikyo (an extreme Japanese religious cult which belongs to category of apocalyptic religious groups) performed a chemical terrorist attack on five subway trains in Tokyo, by releasing sarin, a deadly nerve gas.[17] Eleven victims of the attack died, while up to five thousand were injured, some becoming chronically ill. Further

[16] Clarke, S. C. (2002). Bioterrorism: An overview. *British Journal of Biomedical Science, 59*(4), 232–4. https://search.proquest.com/docview/220149820?accountid=31553

[17] Robery Jay Lifton, *Destroying the World to Save It: Aum Shinrikyo, Apocalyptic Violence, and the New Global Terrorism*, Picador, London 2000.

investigation found that the cult was also in possession of biological agents anthrax and botulinum toxin which it tried to apply in the attacks, but fortunately failed, due to the use of incorrect strains), The group also experimented with Q fever and attempted to acquire the Ebola virus.[18]

In second half of September 2001 (only a week after the famous terrorist attacks in New York and Washington), the "anthrax crisis" shook both the global and American public. Several letters containing anthrax spores were sent via U.S. mail to media offices and high-ranking politicians. The action caused the death of five people and infection of 17 others.[19]

These events were a kind of a crossroad, as before them the specter of bioterrorism could be perceived by many only as a subject of Cold-War fiction, but suddenly it became a terrifying reality, tangible to each individual.[20]

The described cases strongly warned the world about the increased level of the security threat that the weapons of mass destruction (including the biological ones) in hands of non-state actors may represent, especially for communities in large urban areas with high population density.

The risk of the use of biological weapons for this purpose is growing due to:

- Simple production of certain biological agents,
- a wide availability of scientific information through publications and the Internet
- a large number of institutional and non-institutional laboratories (microbiology, molecular biology, genetic) without a complete insight into their operations.

The strongest effects can be caused by large, well-equipped, sometimes state-assisted organizations, able to use modern scientific knowledge, broad arsenal of bioweapons and sophisticated equipment and technology for their production and dissemination. Somewhat smaller effects can be caused by poorly equipped, smaller organizations, and the smallest effect can be caused by small groups or individuals, usually in attempts to assassinate certain persons or to incite panic.

Biological weapons can penetrate the body in three different ways: Inhalation represents the most likely way; inhaling infectious organisms or toxins found in the air. Another way is through ingestion or swallowing, allowing the infection or intoxication through the digestive tract. Absorption through the mucous membrane exposure, through the skin or as a result of wounds or scratches is the third possibility. In addition, biological weapons can damage the material resources or render them unusable.

Particular danger of biological weapons is in their diversity and in the difficulties in assessing the manner in which they would be used, which is the main problem in detecting and responding to such threats, especially when used in undercover and sudden attacks. In addition, bioterrorism can be a powerful factor of destabilization

---

[18] S. E., Meulenbelt, & M. S. Nieuwenhuizen (2015), 838.

[19] Web page of the University of California, Los Angeles (UCLA), American Anthrax Outbreak of 2001: http://www.ph.ucla.edu/epi/bioter/detect/antdetect_intro.html, retrieved 24/06/2017.

[20] Fidler, D. P. (2002). Bioterrorism, public health, and international law. *Chicago Journal of International Law, 3*(1), 7–26.

of a country, especially if it aligns with great powers, as a necessity of most countries in the world that are uncompromising in the combat against terrorism that threatens their very existence. In fact, there is a possibility that in addition to the existing weapons used by terrorists, arsenals of weapons of mass destruction and dangerous materials may get out of control and the international community in such circumstances becomes significantly more susceptible to terrorism. The essential difference between conventional and biological terrorism is in the fact that the conventional means can be controlled in some way, while biological weapons, be they technological, natural or genetically modified agents, once they escape control, become virtually untouchable and unstoppable in their disastrous effects on a wider area or continent.

The need for cooperation between nations in eliminating the threat of biological terrorism in the twenty-first century is more pronounced than ever. The rapid development of science, technology and knowledge brings with it harmful consequences of unprecedented proportions if used, inter alia, for terrorist purposes.

The United Nations General Assembly passed in 2005 the Global strategy to combat terrorism. It emphasizes the importance of addressing the issues that contribute to the manifestation of terrorism, such as unresolved conflicts, discrimination, human rights violations etc. The Strategy established the Working Group on combating terrorism. The role of the working group is to strengthen the coordination and coherence of nations in the fight against terrorism. The key responsibility of the working group is to provide technical assistance to the States in the implementation of the Strategy and measures to prevent the spread of terrorism; measures for preventing and combating terrorism; capacity building for the prevention and fight against terrorism, and for strengthening the role of the UN, measures to ensure respect for human rights and the rule of law. Contemporary risks require a warning system which shall be applicable not only to technical and technological accidents, but also to natural disasters and threats by terrorist attacks. The warning system is very important in saving lives and property.[21]

15.6 Reasons for Strategic Reflection

In the past, the term "strategy" was primarily applied to the military, but in the modern era it has acquired much wider implementation. In the most general sense, strategy is a "long-term planning and political forecasting with a view to ensuring freedom of action, social freedoms, quality of life and order of the state on the basis of the Constitution in order to achieve common political concept".

Considering the strategic approach to biological weapons, we can rightfully say that it is not a phenomenon of the modern era because it existed during the Cold War, where it frequently changed and adapted. In the middle of the last century,

[21] Colonel Katarina Strbac PhD, Emergencies-how to manage them? Institute for Strategic research and The OSCE Mission to Serbia, Belgrade 2009.

biological weapons gained strategic importance in waging modern wars. This paper attaches a special importance to the period at the end of the last century when biological weapons got new contours in the form of bioterrorism.

The strategy for countering bioterrorism is an example of modern approach to the control and prevention of this phenomenon. In addition, the strategy can be seen as an expression of the evolution of the control of biological weapons focusing on the projection of future manifestations of bioterrorism, in order to take optimal measures in countering this phenomenon. The danger of biological weapons is the dark side of globalization, so we often hear appeals for global implementation of prevention and accountability. Preventing the use of biological weapons creates a new chapter for the human race in the form of a long-term fight against the deadly mycobacteria. Bearing in mind the increasing threats of bioterrorism, it is necessary that all countries consider plans of preventive action to keep from possible bioterrorist attacks. All countries should seek to prevent the use of biological weapons, responsibility should therefore be shared, and because of the use of toxic substances aimed at the destruction and endangering people, animals and useful plants, the protocol on the prohibition of poisonous or other gases and bacteriological methods of warfare was signed in Geneva in 1925.

Faced with the possibility of using biological weapons for terrorist purposes, the governments have intensified efforts at the international and national levels aimed primarily at encouraging the introduction and use of a strategic approach in controlling biological weapons in particular in terms of possible misuse for terrorist purposes. The strategic concept is a general and systematic approach to the basic features of bioterrorism in order to make rational use of available resources and more efficient prevention of this phenomenon. In addition, the strategy of countering bioterrorism is linked to the achievement of the strategic objectives pursued by individual countries and the international community as a whole. As a factor that determines the strategic approach, it should be noted that a biological weapon has its advantages over conventional, nuclear or chemical weapons. Therefore, the threat of biological weapons requires a different paradigm than a defensive threat from conventional or other weapons of mass destruction.

Of course, that specific contribution to the strategic orientation is the fact that terrorism is basically is only one of the possible forms of its manifestation.

Bioterrorism as a relatively new phenomenon further adds to the complexity of the fight against terrorism and dealing with its consequences. In fact, this special type of weapons of mass destruction is an increasing and formidable addition to the terrorist arsenal. Its destructive potential is so great that it is now considered a strategic threat to many countries that can cause suffering on a large scale, but also significant political consequences.

The aforementioned indicates that focusing on preventive action in security planning and establishing a long-term strategy to maintain global security, emerge as imperatives. In addition, the phenomenon of terrorism, which will not disappear overnight no matter what measures were taken, demands long-term strategies. However, at the same time, terrorism is a continuous threat that requires constant vigilance, with appropriate and timely measures of detecting and preventing terrorist attacks. Counter-terrorist strategy must involve both emergency and long-term measures and actions, where the basic prerequisite for efficiency of anti-terrorist strategy is a multidisciplinary approach.

15.7 Proliferation Security Initiative (PSI) as an Attempt of Global Approach

In December 2002, the United States of America adopted the National Strategy to Combat Weapons of Mass Destruction (including biological weapons).[22] The timing of the action was closely linked to the dramatic events of the previous year that shook the world and the homeland public in the USA. The terrorist attacks in New York and Washington D.C., as well as the case of the intentional dissemination of Anthrax spores via the national mail system, and use of infected letters imposed new concerns for security and public safety, including public health management.[23] For better understanding of the context and the moment in which the Strategy was adopted, it is necessary to mention the intensive (and lasting) diplomatic campaign against the Iraqi president Saddam Hussein, who was accused of the development of a WMD program (mainly chemical, but potentially also nuclear). Only 4 months later, the military invasion of Iraq started in March 2003.

The National Strategy to Combat Weapons of Mass Destruction identified the need for more robust tools, capable of halting the proliferation of WMD around the world, and specifically identified **interdiction** as an area which requested a particular attention.[24] Soon after the adoption of the National strategy, a global effort of the U.S. diplomacy was launched in order to stop trafficking of weapons of mass destruction, their delivery systems, and related materials to and from the state and non-state actors of proliferation concern. ***The Proliferation Security Initiative* (PSI)**[25] was launched in Krakow (Poland) on May 31, 2003, under the leadership of the president George W. Bush and additionally supported by the next U.S. president Barrack Obama in his Prague speech of April 2009.[26] By 2017, over 100 governments formally endorsed this voluntary initiative aimed at enhancing both the collective and individual capabilities of partner nations to perform timely and appropriate action in response to a fast-changing proliferation threat environment.

The basic idea of the PSI was to serve as a complement to existing counter proliferation efforts, by coordinating activities of participating states, in accordance with national legal frameworks and international law. The declared ambition was to

---

[22] National Strategy to Combat Weapons of Mass Destruction of USA, published September 17, 2002, https://www.armscontrol.org/print/1184

[23] Maddox, P. J. (2001). Bioterrorism: A renewed public health threat. *Dermatology Nursing, 13*(6), 437–41. Retrieved from https://search.proquest.com/docview/224832248?accountid=31553

[24] https://www.state.gov/t/isn/c10390.htm

[25] Official web page of the Proliferation Security Initiative: http://www.psi-online.info

[26] https://www.state.gov/t/isn/c10390.htm

bring together all the states which perceived expansion of WMD as a significant security concern, regardless of their geographic location, size, or diplomatic impact, economic and military strength. The process is based on support to the Statement of Interdiction Principles[27] and readiness of each endorsing nation to cooperate with any state whose ships, flags, ports, territorial waters, airspace, or land might be used for proliferation purposes by states and non-state actors of proliferation concern. During 17 years of its existence, the PSI became an important tool in efforts of the USA to block illegal markets, detect and intercept transit of materials suitable for production of WMD, and use financial tools to harm this trade. It proved to be an innovative and proactive approach to preventing proliferation. The initiative depends on voluntary actions by states that are consistent with their national legal authorities and relevant international law and frameworks. Participants of the PSI use existing authorities (national and international), aiming at suppressing trafficking of material, technology and all the other resources related to WMD.

The goal of the United States is to strengthen and expand the PSI, keeping it as an effective mechanism to stop proliferation of WMD. The efforts of the PSI include the support of diplomatic, financial, military, customs, law enforcement, and other security experts and assets to interdiction exercises, by hosting international meetings, workshops, and exercises and by working with specific partner states to improve their capacity for combating the proliferation of WMD. One of the recent achievements is founding of the Counter WMD (C-WMD) Network, established in 2015 in cooperation with the RACVIAC Centre for Security Cooperation in South-Eastern Europe.

15.7.1 European Approach to Bioterrorism

Bioterrorism was neither a political priority of the European Union, nor a priority of the member-states, before the deliberate anthrax release in the United States of America (September and October 2001). The aforementioned incidents in combination with the terrorist attacks in New York and Washington D.C. completely changed the international perception of the risk of bioterrorism. The agencies responsible for civil protection of the European countries, as well as their security forces were kept in the increased state of preparedness. Many cases of suspicious mail items (containing powders and suspected of being contaminated with anthrax) emerged, which forced the medical institutions to examine them and apply emergency procedures. Although there was no true bioterrorist attack on the European soil, the pressure on national governments and the Union as whole was growing. An efficient response to the new type of threat was requested. Plans for preparedness, response and actions suddenly became a higher priority of the EU member states. Fear of bioterrorism

[27] Statement of Interdiction Principles, presented at the web-page of the U. S. Government: https://www.state.gov/t/isn/c27726.htm

resulted in a debate on the need to reinforce existing public health structures responsible for monitoring and controlling diseases.

The European Commission initiated a number of coordinated actions across the areas of civil protection, healthcare, enterprise (pharmaceuticals), research, nuclear, transport and energy. The Health Security Committee (HSC) was established in 2001, joining representatives of the national Health Ministers, in order to promote cooperation in countering bioterrorism. On November 15th 2001, the Health Council of the EU issued Conclusions, calling on the Commission to develop an Action programme of cooperation on preparedness and response to biological and chemical agent threats.[28] On December 17th 2001, the Health Security Committee agreed on a program for cooperation on preparedness and response to biologic and chemical agent attacks (Health Security Programme). The main goal of the Programme was to improve cooperation between the member states (using assistance of the European Commission) and to facilitate collaboration between stakeholders (national authorities) responsible for preparedness of public health system for bioterrorism. In May 2002, The European Commission's Task Force on Bioterrorism[29] was established, involving nine national experts and six commission officials. The task force was available 24 hours a day, 7 days a week to facilitate the process.

In January 2003, five persons were arrested in the capital of the United Kingdom, on suspicion of conspiracy and the bio-terrorist attack in London underground ("the ricin plot"). The case was later found to be a false alarm (2 years later), but at the moment it was understood as an indicator of the necessity to increase efforts.

On June 2nd 2003, the European Commission issued a Communication to the Council and the European Parliament on cooperation in the European Union on preparedness and response to Biological and Chemical agent attacks (Health security).[30] On the peak of global concerns related to bioterrorism and clandestine programs of WMD in "rogue states", the European Union adopted its first *Strategy against Proliferation of Weapons of Mass destruction* on December 12th 2003.[31] In addition, the European Union adopted the Action Plan on biological and toxin weapons (2006),[32] which obliged the Member States to annually report to the UN on

[28] Eur-lex, access to European Union law, Communication from the Commission of 2 June 2003 to the Council and the European Parliament on cooperation in the European Union on preparedness and response to Biological and Chemical agent attacks (Health security) [COM(2003) 320,final, http://eur-lex.europa.eu/legal-content/EN/TXT/?uri=LEGISSUM:c11576

[29] US National Library of Medicine, National Institute of Health, The European Commission's Task Force on Bioterrorism. https://www.ncbi.nlm.nih.gov/pmc/articles/PMC3033083/

[30] Eur-lex, access to European Union law, Communication from the Commission of 2 June 2003 to the Council and the European Parliament on cooperation in the European Union on preparedness and response to Biological and Chemical agent attacks (Health security) [COM(2003) 320,final, http://eur-lex.europa.eu/legal-content/EN/TXT/?uri=LEGISSUM:c11576

[31] Legal web portal of the European Union, EURLEX: http://eur-lex.europa.eu/legal-content/EN/TXT/?uri=URISERV%3Al33234

[32] EU Action Plan on biological and toxin weapons, complementary to the EU Joint Action in support of the BTWC (2006/C 57/01), http://eur-lex.europa.eu/legal-content/EN/TXT/?uri=uriserv%3AOJ.C_.2006.057.01.0001.01.ENG

the results and confidence building measures. Also, the UN Secretary General is authorized to issue lists of relevant experts and laboratories which might be under investigation of alleged use of chemical or biological weapons. The Strategy *against Proliferation of Weapons of Mass destruction* was further reinforced in 2008 by the *New lines for Action*, aiming at better coordination activities on the level of the Union.[33]

The European Union Non-Proliferation Consortium[34] (a European network of independent non- proliferation think tanks in support of the implementation of the EU strategy against Proliferation of Weapons of Mass Destruction) was established in June 2010. The consortium united the efforts of over 70 foreign policy institutions and research centres from across the EU to encourage political and security-related dialogue and the long-term discussion of measures to combat the proliferation of weapons of mass destruction (WMD) and their delivery systems. Issues of biological weapons and response to bioterrorism are among the main topics researched by the consortium members.[35]

Awareness of the necessity to control biological and other weapons of mass destruction is evident in the EU Global Strategy on Foreign and Security Policy (2016). However, the WMD threat was given more attention in the previous European Security Strategy of 2003. A mitigating factor for that can only be found in the fact that in the meantime the EU *Strategy against Proliferation of WMD* was adopted, with a purpose to strengthen the Convention on the Prohibition of the Development, Production and Stockpiling of Bacteriological (Biological) and Toxic weapons and on their destruction.

The Stockholm International Peace Research Institute (SIPRI) estimates that in 2016 many other issues dominated the agenda, while non-proliferation and matters of arms control "were not given a prominent place among the priorities of the Global Strategy".[36] The institute further suggests that "one or more new strategy documents are required and, in this context, the EU should also pursue WMD-related contingency planning to increase preparedness and prevent or counter crises".

The importance of the problem of proliferation of weapons of mass destruction is furthered by the fact that the proliferation of WMD is closely connected with other global risks such as organized crime, international terrorism, regional conflicts and other global security challenges, which are conducive to the proliferation of WMDs.

---

[33] EU New Lines for Action, http://register.consilium.europa.eu/doc/srv?l=EN&f=ST%20 17172%202008%20INIT

[34] EU non-proliferation consortium, The European Network of Independent Non-proliferation Think Tanks https://www.nonproliferation.eu

[35] EU non-proliferation consortium, The European Network of Independent Non-proliferation Think Tanks https://www.nonproliferation.eu/thematics/biological-weapons

[36] SIPRI, The European Union and weapons of mass destruction: A follow-on to the global strategy? https://www.sipri.org/publications/2017/eu-non-proliferation-papers/european-union-and-weapons-mass-destruction

The current migration crisis also raises an issue of possibility of aggressive biological agents reaching parts of Europe in which they have not been present before. Cooperation on national and international levels of all agencies involved in the combat against organized crime must be stronger, accepting standards and procedures which will strengthen governments and EU abilities to confront one of the most visible asymmetric threats to security of the entire European continent.[37]

Bearing in mind the dynamic nature of modern threats and that defence no longer exists in archaic terms, it is necessary to involve all Member States, the candidate countries, including the countries that have just started the process of accession, such as the Republic of Serbia[38] in establishment of common views and strategic approach to the problem of terrorism in Europe.

Analyzing the nature of the risks, threats and dangers to the European Security Strategy European security, it is faced with the following:

- proliferation of weapons of mass destruction, especially in combination with international terrorism,
- terrorism especially on a large scale ("super terrorism", "hyper terrorism," "mega terrorism"),

and

- regional conflicts that occur as sources of other threats such as terrorism, proliferation of WMD, organized crime and extremism.

Although majority of the EU member states are primarily focused on issues relevant to their own security, the concern over proliferation of weapons of mass destruction and international terrorism (which might obtain and apply nonconventional weaponry, including biological agents) remains in most of the national security strategies.

15.8 Bioterrorism in National Security Strategies

Because of existing risks as well as other characteristics that set it apart from other weapons, preventing the use of biological weapons for terrorist purposes is a top priority of preserving national security of modern states. Many of the strategies of the national security state pay special attention to this phenomenon. Very often, the biological weapons are viewed in the context of weapons of mass destruction and

[37] Strbac Katarina, Branislav Milosavljevic, Boban Radivojevic, Some Aspects of Illegal Migrations, Zborník príspevkov6. medzinárodnej vedeckej konferencie, Bezpečné Slovensko A Európska Únia, Vysoká Škola Bezpečnostného Manažérstva V Košiciach, 2012.

[38] Strbac Katarina, The Perspective on Challenges and Complementarities of the Standpoints of the Republic of Serbia and the EU, Proceedings "Security and Defence aspects of the Republic of Serbia's accession to the European union", Strategic Research Institute and OSCE Mission to Serbia, Belgrade, 2010.

separately as it is the case in the US National Security Strategy. In the part relating to the prevention of the spread and use of weapons of mass destruction, in particular, it focuses on the ability of state and non-state actors to procure or develop inter alia, biological weapons, which of course requires adequate response from relevant state entities.

In addition to the current US strategy in the part related to health security, there is a special mention of biological weapons, where it is stressed that the spread of communicable diseases poses an increasing risk despite the scientific and technological advances in their prevention. In particular, in the statement regarding a lack of the capacity to prevent, detect and respond in the event of an outbreak of these diseases. As a world leader in the fight against the current pandemic, the US continues to strengthen the capacity for adequate response capacity and crisis management caused by infectious diseases, which among other things requires the expansion of cooperation through the Global Health program to achieve a safer world and less vulnerable to infectious diseases.

Bright examples are National Security Strategies of Austria and Bulgaria, which similarly assess the potential asymmetric threats and particularly terrorism and the proliferation of weapons of mass destruction. This is the case with the strategies of other countries with similar approach so we can give a general conclusion. In addition, countries pay special attention to the proliferation of weapons for mass destruction because they can be threatened with such weapons or their territory might be used for transit.

Faced with the threat of biological terrorism and its possible consequences, many states have tackled this problem with a lot of attention. Provisions given in national security strategies are implemented through a separate strategy as a general framework for action by all relevant government bodies. The relevant strategies are commonly referred to as strategies for prevention the proliferation of weapons of mass destruction that could be seen as guidelines for improving the coordination and activities at the national, but also at the international level. In addition, the strategy can be seen as a response to the commitments of the countries signatories to UN Security Council Resolution 1540, which was adopted in 2004, which calls all States, in accordance with their national legislation and international law, to undertake joint measures and activities to prevent the spread of weapons of mass destruction and respect international legal instruments.

Strategies are the basis for joint and coordinated action by the state authorities, as well as the continuous improvement and finding mechanisms for the control and prevention of proliferation of biological weapons. As a conclusion, strategies have indicated establishment of specialized authorities for efficient implementation. Adequately formulated strategies enable prevention as one of the key areas to counter the spread of weapons of mass destruction. In this regard, there is a need to strengthen the national capacity of all institutions responsible for the implementation of the Strategy.

15.9 Regional Initiative in South-Eastern Europe

South-Eastern Europe is located in the area intersected by land and maritime smuggling and human trafficking routes from Asia, Africa and Eastern Europe leading towards the Western Europe. Such position makes it particularly sensitive to possibility of becoming a transit area over which material, technologies and qualified or indoctrinated individuals may reach western capitals and commit terrorist attacks, including ones based on application of bio-agents.

The ongoing "European migrant crisis" which started in 2015 brought additional arguments for concern over possible import of disease causing pathogens. Infectious diseases might hit parts of European population, either as a result of inadequate health control of the migrating groups and individuals or due to the intention of terrorist networks to use aggressive micro-organisms as weaponry. Capabilities of individual states to manage such crises could be limited, especially if there is a lack of medicaments and vaccines on stocks for diseases which have been considered eradicated for many decades. The new scope of challenges imposes the need for an increased level of regional cooperation in all phases, including risk analysis, exchange of intelligence, joint planning, capability building, harmonization of strategic and legal frameworks and finally, joint operation.

Among the states of South-Eastern Europe, Croatia was the first to adopt a separate and comprehensive *National Strategy for the Non-Proliferation of Weapons of Mass Destruction* (with the Implementation Plan included),[39] in 2010, one year after achieving full NATO membership and 3 years before its accession to the European Union. That was a logical step, aiming to demonstrate adherence to the Euro-Atlantic perception of key security challenges, as well as a capability to share goals and values with main foreign partners. Previously, Croatia endorsed the Proliferation Security Initiative (PSI) in 2004 and within its framework signed the bilateral Ship Boarding Agreement with the USA.

Development and implementation of such comprehensive strategy which became an integral part of the national crisis management system, required a complex and permanent inter-agency cooperation, coordinated by the Ministry of Foreign and European Affairs, but also with a significant contribution of Ministry of Defense and many other relevant agencies.

As a result of the obtained experience and ambition of Croatia to be facilitator of further regional security integrations, the project "C-WMD Network" was established in 2015, and coordinated by RACVIAC – Centre for Security Cooperation.[40] The project has been supported by PSI, United States European Command (US EUCOM) and Defense Threat Reduction Agency (DTRA). It also enjoys expert

[39] Government of the Republic of Croatia, https://vlada.gov.hr/UserDocsImages//Sjednice/Arhiva//71.%20-%206.pdfhttps://www.google.rs/url?sa=t&rct=j&q=&esrc=s&source=web&cd=1&cad=rja&uact=8&ved=0ahUKEwj5k7eMmNfUAhVI0xoKHc_VDbsQFggkMAA&url=http%3A%2F%2Fwww.un.org%2Fen%2Fsc%2F1540%2Fdocuments%2FCroatia-action-plan.pdf&usg=AFQjCNFy5QbNjPBwhPx3QMsrIr9h3I2M8Q

[40] RACVIAC: www.racviac.org

support of the relevant European institutions – stakeholders in the process of WMD proliferation control.

As a result of the project, Montenegro was the second state of the region to adopt the Strategy for Non-Proliferation of Weapons of Mass Destruction (2016–2020), in September 2016. Most of the remaining countries of the region have made a political decision to develop similar strategy, based on the pattern provided by PSI experts and RACVIAC. By June 2017, the list of governments who declared the intention to adopt the strategy included Albania, Bulgaria, Macedonia, Romania and Moldova. Serbia was still seeking consensus over the issue whether the existing strategic framework should be considered as sufficient or an additional "roof document" should be added, in order to bind all strategies. Ukraine was participating in the process, without declared obligation to adopt the strategy, but willing to use the obtained information for improvement of its ongoing practice. Although Kosovo is considered by Serbia as its inseparable constitutional part, it participated in the regional process, under the terms of the Brussels Process and started working on development of the Strategy. In the end, all regional countries individually, but in cooperation with each other, should develop new security culture. Such security culture should minimize the influence of the past negative experiences, prejudices, and stereotypes, thus making the regional security one of the key factors in overall regional development.[41]

15.10 National Approach Case Serbia

Unlike Croatia and Montenegro, Republic of Serbia has not yet developed a Strategy for prevention of proliferation of weapons of mass destruction. On the other hand, Serbia has ratified many conventions, among others, the Convention on the Prohibition of the Development, Production and Stockpiling of Bacteriological (Biological) and Toxic Weapons and on their Destruction. As a UN member state the Republic of Serbia has ratified the UN Security Council Resolution 1540 concerning the proliferation of weapons of mass destruction and their transmission and thus defines the obligations assumed in accordance with the aforementioned resolution.

Within the negotiation process on Serbia's accession to the European Union, the Negotiating Group for the Chapter 31 (Common Foreign Security and Defence Policy) came to the conclusion that a comprehensive Strategy on Non-Proliferation of WMD should be adopted. The Ministry of Foreign Affairs initiated the process in second half of 2016, by requesting opinion of all the relevant governmental agencies and suggesting that a positive approach would be useful. However, even a year later, the consensus has not been achieved yet. Some agencies consider that efforts to develop such strategy would be a double and redundant effort. Their argument is

[41] Strbac Katarina, Miroslav Mitrovic, Asymmetric threats-common response in Western Balkans, The Review of international Affairs, Belgrade, 2011.

that most of the content that should be a part of such strategy already exists in other strategic documents covering issues of counter-terrorism and CBRN. In the moment when this paper is being submitted, it is not clear yet whether Serbia is going to adopt the Strategy or the existing framework will be considered sufficient.

Having in mind geographical position of the Republic of Serbia, the security of the country may be burdened by the crisis in the immediate neighborhood, but also in the wider region, especially in the area that includes the Middle East, the Caucasus, North Africa and Mediterranean. All the mentioned areas are unstable in the security area and have manifestations of transnational threats to security and their transfer to the European continent. Further, Serbia is on the transit route with intersecting smuggling routes from Asia, Africa and Eastern Europe to Western Europe, which makes it particularly exposed to the possibility of smuggling of biological weapons.

In addition, the possibility of proliferation of this weapon and delivery and obtaining them from non-state actors, especially terrorists, represent a serious threat because there is a real risk of the spread of technology and information for the use of biological weapons in actions that would lead to killing and destruction of large proportions. Such events would take place outside the existing control regime. There is also the danger that terrorist groups exploit the migrant routes for the proliferation of weapons and the perpetration of terrorist attacks. Based on these reasons, there are visible arguments why a special strategy is needed, to encompass this area and formulate three goals:

- Prevent possibility of individuals, groups or states to achieve illegal possession of weapons of mass destruction,
- Prevent the use of WMD by criminal and terrorist entities and
- Eliminate and reduce risks in case of possible use of weapons of mass destruction.

In order for these objectives to be achieved, it is necessary to create adequate security conditions and actively participate in the achievement of international cooperation in this field. Successful implementation of the Strategy implies the incorporation of the planned objectives and measures for its implementation to other strategic and planning documents and procedures for the adoption of national policies in the security sector. Strategic planning at the national level involves building up a comprehensive policy to prevent bioterrorism. The necessity of building a national strategy is particularly important because it is a specific form of endangering national security, which requires the involvement of a large number of actors both in the process of preventing and eliminating the consequences. For the process of creating national strategies, it is of crucial importance to conduct real consideration of the scope of the threat of biological terrorism both at the national and international levels.

An objective view of the substance of the existing and the possibility of future development and emergence of a form of biological terrorism constitutes a sufficient reason to follow appropriate actions of the state and of all relevant actors in its eradication.

In addition, strategy must have an action plan which should enable the initial work program for the practical implementation of the basic principles. The implementation of a non-proliferation of WMD strategy requires time and is therefore essential to have action plans containing urgent and long term measures. Measures for immediate action include implementation time required, political and legal instruments for the implementation, and the expected costs of implementation. However, there is not a political instrument, no "magic wand" that can solve the problem, but there is the need for integration of multi-functioning strategy and cooperation at the international level.

Finally, looking globally, the European and regional security environment mostly depends on capabilities to positively direct political and security processes in this area. It seems that in spite of all efforts made so far, Balkan countries have to do much more in the field of protection from asymmetric threats than they do today.[42]

15.11 Conclusion

Biological terrorism as a phenomenon of modern times shows unpredictability, fanaticism and cruelty which might be considered as significant threats to humanity today. Bioterrorism can be a powerful factor in the destabilization of a country, therefore it needs to be considered very seriously. Despite the fact that the NBC terrorism was dominant throughout the twentieth century, it is certainly the main danger and a threat to humanity in the twenty-first century as well. Bioterrorism is a specific security threat because it is characterized by a combination of high mortality rates, relatively simple method of production and the possibility of covert use.

Simplicity of the misuse of biological weapons is perhaps best demonstrated by its definition as "the atomic bomb of the poor" because of the relatively low cost of production. From the standpoint of terrorist organizations and groups, the use of biological weapons brings more advantages over the conventional explosive materials. Biological weapons produce a high level of mortality of humans, animals and plants, very small amounts of pathogens can achieve a high degree of destruction and they are relatively easily and quickly activated (released).The possibility of permanent activation of the equipment that is necessary for production is inexpensive and easily available. The trend of increasing casualties in terrorist attacks in recent years suggests that terrorists are looking for new strategies, methods, weapons and funds to make the effects of their attacks as large as possible. Plenty of evidence suggest that the use of nuclear, radiological, and most of all biological and chemical weapons is likely and that we should be prepared for such scenarios. Biological "war" is quite possible, if not already our reality. The fact is that after the spread of the contaminated letters in the US a biological war had officially started. Those who meet it unprepared will face unforeseeable consequences. It is essential

[42] Strbac Katarina, Evolving Asymmetric Threats in the Balkans, NATO Science for Peace and Security Series E: Human and Societal Dynamics-Vol 85, 2010.

that we undertake a number of measures, means and procedures in order to have, as much as possible, safe and bright future and reduce potential threats to the lowest possible level.[43]

Literature

1. Chevrier MI, Chomiczewski K, Garrigue H (eds) (2004). The implementation of legally binding measures to strengthen the biological and toxin weapons convention: proceedings of the NATO Advanced Study Institute, Held in Budapest, Hungary, 2001. Volume 150 of NATO science series: Mathematics, physics, and chemistry, illustrated edn. Springer
2. Clarke SC (2002) Bioterrorism: an overview. Br J Biomed Sci 59(4):232–234
3. Strbac K (2009) Emergencies-how to manage them? Institute for Strategic research and The OSCE Mission to Serbia, Belgrade
4. Costa H, Baños J (2016) Bioterrorism in the literature of the nineteenth century: the case of wells and the stolen bacillus. Cogent Arts & Humanit 3(1):1224538
5. Encyclopaedia Britannica. League of nations. Internet edn
6. Meier E-C, Nelte K-M, Schaefer H-U (2006) Woerterbuch yur Sichereitspolitik-Deutschland in einem veraenderten internationalen Umfeld. 6. vollstaending ueberarbeitete Auflage. Verlag E.S. Mittler&Sohn, Hamburg
7. EU New Lines for Action. http://register.consilium.europa.eu/doc/srv?l=EN&f=ST%20 17172%202008%20INIT
8. European security strategy 2008. http://www.isac-fund.org/download/Evropska%20strategija%20bezbednosti.pdf
9. Fidler DP (2002) Bioterrorism, public health, and international law. Chicago J Int Law 3(1):7–26
10. Foster T (2006) George, focus on bioterrorism. Nova Science Publishers, London
11. Gaćinović R (2005) Terorizam. Draslar, Beograd
12. Legal web portal of the European Union, EURLEX. http://eur-lex.europa.eu/legal-content/EN/TXT/?uri=URISERV%3Al33234
13. EU Action Plan on biological and toxin weapons, complementary to the EU Joint Action in support of the BTWC (2006/C 57/01). http://eur-lex.europa.eu/legal-content/EN/TXT/?uri=uriserv%3AOJ.C_.2006.057.01.0001.01.ENG
14. EU non-proliferation consortium, The European Network of Independent Non-proliferation Think Tanks. https://www.nonproliferation.eu
15. EU non-proliferation consortium, The European Network of Independent Non-proliferation Think Tanks. https://www.nonproliferation.eu/thematics/biological-weapons
16. SIPRI, The European Union and weapons of mass destruction: A follow-on to the global strategy?
17. https://ihldatabases.icrc.org/applic/ihl/ihl.nsf/Article.xsp?action=openDocument&documentId=8FCF14D950797012C12563CD00515C0A
18. International Committee of the Red Cross. https://ihl-databases.icrc.org/ihl/INTRO/145?OpenDocument
19. https://search.proquest.com/docview/220149820?accountid=31553
20. https://search.proquest.com/docview/237212498?accountid=31553
21. Government of the Republic of Croatia. https://vlada.gov.hr/UserDocsImages//Sjednice/Arhiva//71.%20-%206.pdfhttps://www.google.rs/url?sa=t&rct=j&q=&esrc=s&source=web&

[43] Strbac Katarina, Welcome address: Implications of Climate Change and Disasters on Military Activities NATO Science for Peace and Security Series – C: Environmental Security, Springer, 2016.

cd=1&cad=rja&uact=8&ved=0ahUKEwj5k7eMmNfUAhVI0xoKHc_VDbsQFggkMAA&url=http%3A%2F%2Fwww.un.org%2Fen%2Fsc%2F1540%2Fdocuments%2FCroatia-action-plan.pdf&usg=AFQjCNFy5QbNjPBwhPx3QMsrIr9h3I2M8Q
22. Arms Control Association. https://www.armscontrol.org/factsheets/cbwprolif. Chemical and Biological weapon at glance
23. Government of the Republic of Bulgaria. https://www.bbn.gov.pl/ftp/dok/07/BGR_National_Security_Strategy_Republic_Bulgaria_2011.pdf
24. Government of the Republic of Austria. https://www.bka.gv.at/DocView.axd?CobId=52251
25. https://www.britannica.com/topic/League-of-Nations (30/06/2017)
26. US National Library of Medicine, National Institute of Health, The European Commission's Task Force on Bioterrorism. https://www.ncbi.nlm.nih.gov/pmc/articles/PMC3033083/https://www.peacepalacelibrary.nl/2016/04/preventing-bioterrorism-risk-and-legal-instruments/
27. US Department of State. Proliferation Security initiative. https://www.state.gov/t/isn/c10390.htm
28. Organization of United Nations. The Biological weapon convention. https://www.un.org/disarmament/geneva/bwc/
29. Organization of United Nations, 1925 Geneva Protocol. https://www.un.org/disarmament/wmd/bio/1925-geneva-protocol
30. International Committee of the Red Cross web database. https://ihl-databases.icrc.org/ihl/INTRO/150?OpenDocument
31. International Convention for the Suppression of the financing of terrorism (1999)
32. International Convention for the Suppression of the terrorist bombings (1997)
33. Jončić V (2010) Međunarodno humanitarno pravo. Pravni fakultet, Beograd
34. Jović R, Savić A (2004) Bioterorizam, biološki rat. Institut za političke studije, Beograd
35. Larsen MR (2010) Al qaeda weapons of mass destruction threat:hype or reality. Belfer center for science and international affairs, Cambridge. http://www.belfercenter.org/sites/default/files/files/publication/al-qaeda-wmd-threat.pdf
36. Legal web portal of the European Union, EURLEX
37. Maddox PJ (2001) Bioterrorism: a renewed public health threat. Dermatol Nurs 13(6):437–441. https://search.proquest.com/docview/224832248?accountid=31553
38. Mainuddin RG (2014) Prohibiting chemical and biological weapons: multilateral regimes and their evolution. Choice 51(12):2265–2266. https://search.proquest.com/docview/1548715180?accountid=31553
39. Meulenbelt SE, Nieuwenhuizen MS (2015) Non-state actors' pursuit of CBRN weapons: from motivation to potential humanitarian consequences. International Review
40. National Security Strategy USA. http://nssarchive.us/national-security-strategy-2015/
41. NATO Glossary of terms and definitions (AAP-6). http://www.dtic.mil/doctrine/doctrine/other/aap6.pdf
42. Meulenbelt ES, Nieuwenhuizen SM (2015) The human cost of nuclear weapons, Nonstate actors pursuit of CBRN weapons:From motivation to humanitarian consequences. Int Rev Red Cross 97(899):831–858
43. Official web page of the Proliferation Security Initiative. http://www.psi-online.info
44. Official web-site of the RACVIAC. www.racviac.org
45. Ostfield ML (2004) Bioterrorism as a foreign policy issue. SAIS Rev Int Aff 24(1):131–146. https://search.proquest.com/docview/231316387?accountid=31553
46. Paulun M (2003) Weapons of mass destruction, the first responder. Random, New York
47. Lifton RJ (2000) Destroying the world to save it: Aum Shinrikyo, apocalyptic violence, and the new global terrorism. Picador, London
48. Statement of Interdiction Principles. Presented at the web-page of the U. S. Government. https://www.state.gov/t/isn/c27726.htm
49. Katarina S, Milosavljevic B, Radivojevic B (2012) Some Aspects of Illegal Migrations, Zborník príspevkov6. medzinárodnej vedeckej konferencie, Bezpečné Slovensko A Európska Únia, Vysoká Škola Bezpečnostného Manažérstva V Košiciach

50. Strbac Katarina (2010) Evolving asymmetric threats in the Balkans. NATO Science for Peace and security series E: human and societal dynamics, vol 85. IOS Press, Amsterdam
51. Katarina S, Mitrovic M (2011) Asymmetric threats-common response in Western Balkans. The Review of International Affairs, Belgrade
52. Katarina S (2010) The perspective on challenges and complementarities of the standpoints of the Republic of Serbia and the EU. In: Proceedings "Security and Defence aspects of the Republic of Serbia's accession to the European union". Strategic Research Institute and OSCE Mission to Serbia, Belgrade
53. Katarina S (2016) Welcome address: implications of Climate Change and disasters on Military Activities. NATO Science for peace and security series – C: environmental security. Springer
54. The Final Protocol of the First Geneva Conference (1899)
55. United Nations Office for Disarmament Affairs (UNODA) official web site. https://www.un.org/disarmament/wmd/bio
56. United Nations Security Council Resolution 1540 (2004)
57. Web page of the University of California, Los Angeles (UCLA), American Anthrax Outbreak of 2001. http://www.ph.ucla.edu/epi/bioter/detect/antdetect_intro.html

Chapter 16
Food and Bioterrorism – The Case of Airline Catering

Ines Banjari

Food is essential to life, therefore make it good.

S. Truett Cathy

Abstract Food is the essence of life but food is also at the top of the list of potential means for a bioterroristic attack. Today, more than ever, the threat to the food supply chain seem more vivid. But what have we learned from several well-documented historical examples of the bioterroristic attacks via food? 9/11 has changed how we perceive security, safety and our daily routine. This was the turning point. We have developed and effectively implemented a number of hygienic measures to ensure the safety of the food supply chain, reaching the top level of food security with the introduction of the food defence in 2002. Still, this is a somewhat new concept in many of the world's countries. Some branches specifically adapted the existing standards to fit their needs. Airline catering is one of them. Airline catering is probably one of the most complex operational systems in the world. In light of the constant increase of a number of passengers and flights operated every year, more focus should be put on food security on-board. Food handlers, both ground and on-board staff represent the basis of the food security on-board. Investing in people (continuous education) should be accompanied with procedures that could be introduced to the existing system to improve the security and safety of everyone, not only people on-board but also those on ground.

I. Banjari (✉)
Faculty of Food Technology, Josip Juraj Strossmayer University of Osijek, Osijek, Croatia
e-mail: ibanjari@ptfos.hr

V. Radosavljevic et al. (eds.), *Defence Against Bioterrorism*, NATO Science for Peace and Security Series A: Chemistry and Biology,
https://doi.org/10.1007/978-94-024-1263-5_16

16.1 Introduction

Various people associate different attributes to food. Still, all have one thing in common; food is the essence of life, but we rarely think of it in the context of life-threatening. An average person consumes 25 tons of food in a lifetime, and that fact only places food on the top of the list for a potential bioterroristic attack. Food has been used as mean for the biterorristic attacks [7, 4, 19], and one of the most famous examples was the one from 1984 in Dalles, Oregon by the Rajneeshee group [7, 4, 19].

A wide range of contaminants can be used in bioterroristic attack by food and/or water. More than 180 pathogens have been reported to be potential agents for bioterrorism [3]. If we add up thousands of chemical compounds [24, 25], both synthetic and plant derived (available in databases like *TOXNET* by NIH or *OpenFoodTox* by EFSA) we end up with a threat that cannot and must not be neglected [16]. The ideal threat agent is characterized by the following descriptors:

- Inexpensive and easy to produce
- Highly lethal or infectious
- Resistant to environmental factors
- No effective treatment available
- Low infectious dose
- Transmitted via air, water and food
- Transmitted person-to-person

9/11 has changed how we perceive security, safety and our daily routine. Those events affected people living in almost all parts of the world. Today, terrorism is considered to be the new plague. Truly, Global Terrorism Database confirms that. The turning point was 2001, and again 2004, after that we see a slow but constant increase in the number of terroristic incidents over time. Even though weapons of mass destruction represent the highest concern at the moment, Food and Water Supply has been targeted 302 times so far [13]. According to the Food Adulteration Incidents Registry (FAIR) developed by the Food Protection and Defence Institute, A Homeland Security's Centre of Excellence share of intentional distribution of a contaminated food product is less than 10% of all cases [9]. When speaking about malicious food supply contamination, the four main motivation drivers are behind it: disgruntled employee, industrial sabotage, bioterrorism and Economically Motivated Adulteration (EMA). To get the sense of the context, between 1950 and 2005 [7] there has been 41 bioterrporistic incidents in comparison to more than 200 EMAs since 1980 (FPDI 2012).

Systematic review conducted by Brainard and Hunter [4] contains updated information on malicious attacks on food and water supply from 1946 to mid 2015, also providing motive(s) for the attack. They listed a total of 224 food attacks with a total of 1171 death. While for the majority of these attacks motives were unclear (25%), financial extortion (22%) and political motives (16%) are the most common. Food attacks were located mainly in the United States or Canada (24%), the People's Republic of China (19.6%), while 10.3% in the UK and 11.6% elsewhere in Europe.

16.2 Food Defence

The Global Food System is very complex and represents one of the critical elements that a society depends upon and therefore needs to be maintained and protected [2]. Still, it is so delicate and could be breached easily [16]. Food system can be described as a balance between supply and demand, with consumers being the main driver of the system, regardless of the food industry in focus. The industry developed and effectively implemented a number of food safety standards and some of the globally recognised certification schemes are Hazard Analysis and Critical Control Point (HACCP), International Food Standard (IFS), British Retail Consortium (BRC), the National Sanitation Foundation (NSF), Foundation for Food Safety Certification (FSSC 22000) based on the ISO 22000 and ISO 22002-1 and other.

Until 2001, food safety standards were considered to be sufficient in preserving and protecting our food supply. In 2002 the Public Health Security and Bioterrorism Preparedness and Response Act of 2002 [8] has been published. Its primary objective is the protection of food supply from intentional contamination, and food defence was introduced.

Food defence is a term that involves the active effort to protect the food supply from deliberate contamination that is meant to intentionally harm individuals or organizations [2]. Wherever intentional contamination and food fraud pose a serious threat to the consumers' and public health or business, food defence should be implemented [6]. In other words, intentional contamination has the potential to cause significant public health consequences, devastating economic impacts, loss of public confidence in the safety of food, effectiveness of government, and food insecurity [4].

Food defence is usually built up, integrated into an existing food safety system of a company. Even though both use the same or similar tools and methods, potential consequences are drastically different. The levels of Food Defence are shown in Fig. 16.1.

Food defence plan can be a basic or enhanced, and its complexity is mainly determined by the size of a company. Food defence plans, like food safety plans have four components that involve the development, implementation, testing, and review of the plan. Verification of a particular food defence plan lies in testing, and effective corrective measures need to be implemented accordingly [2, 6]. Vulnerability assessment and mitigation strategies are integrated only into an enhanced food defence plan. Food defence involves documentation and written procedures of the existing food safety system, traceability, documentation and records of the management system, corrective and preventive measures, internal audits, employee training, etc. It is mainly implemented in agricultural production, processing, storage and transport, wholesale and retail distribution and tracing systems and recalls (traceability being one of the obligatory requirements) [23]. It is important to note that the analysis of documented bioterroristic attacks via food and/or water showed that the attacks that targeted raw materials (the field), manufacturers, and retailers were less successful which reflects the high security standards that the food industry have been embraced or have been required to implement [4].

Even though the concept of food defence is relatively new to both the EU and non-EU countries, the most important food safety standards in Europe, BRC and

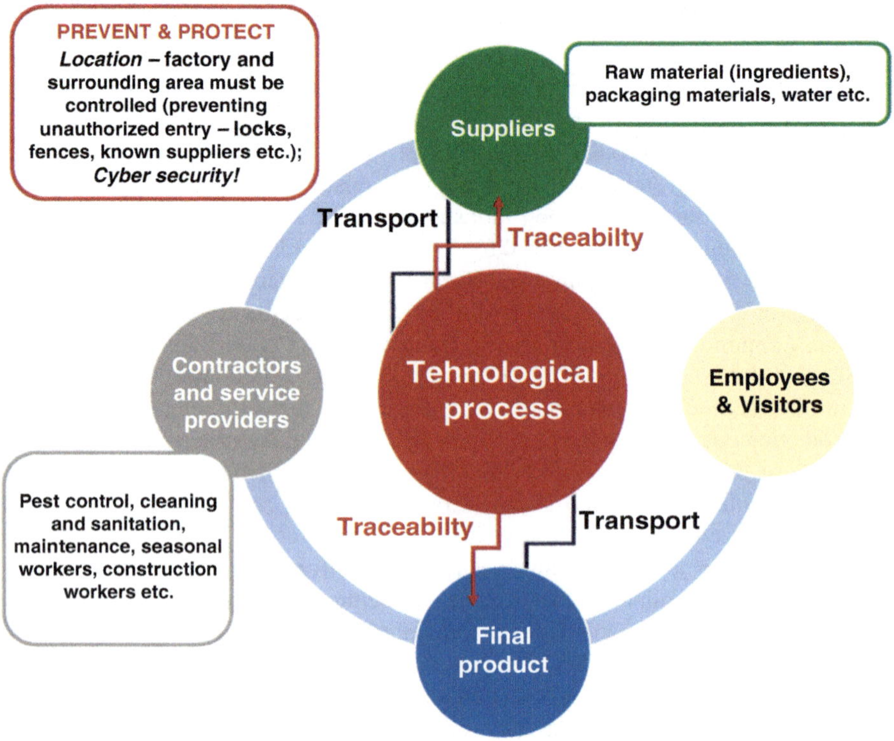

Fig. 16.1 The levels of food defence

IFS introduced food defence requirements as mandatory. These requirements include the implementation of hazard analysis, the assessment of related risks and the identification of critical areas within the subject assessment [5]. Today, more and more attention is given to the IT sector, specifically developed surveillance systems and other cyber security methods [18].

The highest risk for a bioterroristic attack is actually in the final stage, consumption of the food. At this stage, food industry has no control over the food they distributed on the market. Food handlers take over and yet they have been consistently proven as the weakest loop in the food supply chain [1, 20, 21, 26, 28].

16.3 Airline Catering

Each day anywhere between 78,792 and 88,101 flights are operated, having on-board more than 9 million people which is an increase of 48.6% in comparison to year 2000 [27]. The majority of these flights serve food on-board.

Airlines today are mainly concentrated in one of the three large alliances (Fig. 16.2). The biggest one, **Star Alliance** was founded in 1997 and has 27 mem-

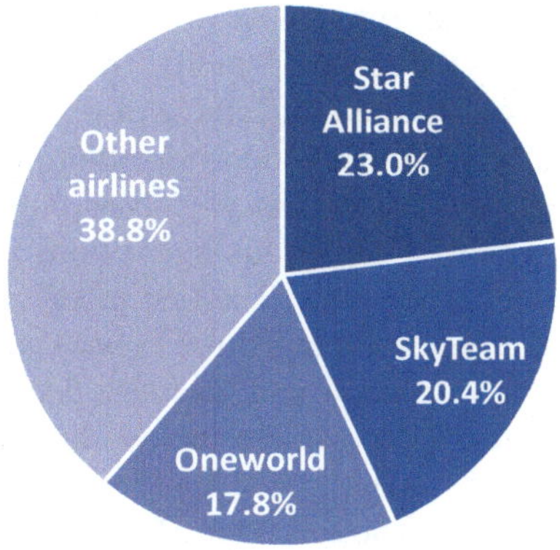

Fig. 16.2 Major airline alliances (prepared according to the World Bank Data [27])

bers including United Airlines, Lufthansa, Air Canada, Air China, and Air India. **Oneworld** was founded in 1999 with 14 members including American Airlines, British Airways, Malaysia Airlines, and Japan Airlines. The youngest, holding the second largest share is **SkyTeam** which was founded in 2000 and has 20 members including Delta Air Lines, Aeroflot, KLM, Aeromexico, and China Airlines.

The first airline meals were served by Handley Page Transport to serve the London–Paris route in October 1919 and passengers could choose from a selection of sandwiches and fruits. Today, the food served on-board varies significantly, from pre-packed food, a picnic bag to *a la carte* meals served for the first class flights. Food on-board can be observed as a marketing tool, but more importantly it usually determines the final impression of a passenger while not being important factor in terms of choosing an airline [17].

Airline catering is probably one of the most complex operational systems in the world [17]. Caterers constantly need to balance between time, weight, waste, quality of the food, innovation and safety. Safety of the food served on-boar is handled by the International Flight Services Association, i.e. the World Food Safety standard for airline caterers (IFSA 2016).

Flight catering is 70% logistics and 30% cooking[1] illustrates the complexity of the airline catering logistics that resembles military logistics and distribution systems.

However, when discussing food on-board from the aspect of a bioterroristic attack, one must consider both food and passenger's characteristics. Passenger's appetite and behaviour prior and during the flight change. The stress at the airport (caused by security checks, various controls, etc.) alter the appetite in a way that a passenger does not think about the food until finally on-board. When finally on-board, passenger finally relaxes, and sometimes, usually because of delays passengers are hungry so the food served is considered as an award, comfort.

Also, passenger's sensory abilities such as smell, sight, and taste are affected by the relatively low humidity and air pressure experienced at altitude. This affects

[1] Quote by he President of KLM Catering (source Jones [17]).

taste buds (up to 30% lowered function) and mucous membranes in the nose (which blunts the sense of smell). Therefore, food served on board is often more seasoned. Due to limited movement of passengers, meals provided must be easily digestible. Alcohol consumption is another important factor. Besides dehydrating the body its effects are more quickly observed in a pressurised cabin [17].

Despite numerous innovations in the supply chain since the 1990s, three basic models pervaded. The first model, in-house catering originates from European airlines who developed their own caterings. With time, the majority were closed, but two, Lufthansa and Swissair decided to further develop their catering divisions. Today, these two are the two largest flight caterings in the world; LSG Sky Chefs serving annually 405 million meals and Gate Gourmet serving 200 million meals annually. The second model is outsourced catering, originating from North America. Here, caterers are responsible for supplying only fresh items, i.e. meals, while for other goods that do not need preparation, packaging or tray assembly (like beverages and duty free goods) airlines negotiated directly with suppliers. More recent model is "buy-on-board", characteristic for low-cost airlines and also used as a business strategy, to reduce flight prices [17]. Still, many airlines due to economic crisis in 2008 decided to adopt off on-flight food completely or went for "buy-on-board".

Caterers have two main roles: preparing items for loading on board and to assemble trays and trolleys [17]. Food products delivered by caterers must be consistent, health safe and in line with the newest trends (innovation in the sector) while relying on just-in-time delivery and production. Even though many steps in the system (Fig. 16.3) can be outsourced, caterers are always responsible for loading and unloading aircraft, recycling and waste disposal because no other stakeholder has the necessary infrastructure (high-loader trucks).

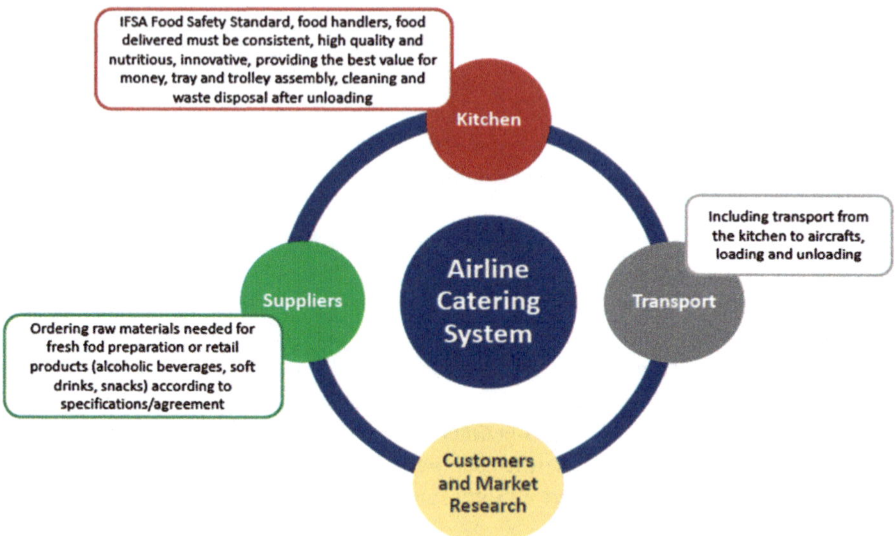

Fig. 16.3 Scheme of the airline catering system (prepared according to Jones [17])

From the aspect of developing an effective food defence plan for the airline catering, considerable attention should be given to the following components: transportation to the aircraft, loading of the food on-board and food handler's behaviour both in the kitchen and on-board.

Flight kitchens are always located near to major airports because of the cost of space and the cost of labour [17]. Therefore, the time needed for delivery is extended. Transportation is usually carried out by using specialist high-loader trucks loaded with trolleys (containing food served on trays). In case of a delay, loading is prolonged and sometimes, food has been already deployed from the kitchen to the loading/unloading area. In conclusion, the average number of people coming in contact with the food prior loading should be minimized as much possible. Additionally, trolleys with electronic (coded) locks cold be used as well, especially in case of a delay (at first they were introduced to prevent theft).

Poor knowledge on food safety among food handlers has been mentioned previously, but two additional aspects must be considered in the context of bioterrorism. Kitchen staff has the lowest wages on the market [15], especially assistant kitchen staff, many of whom have different educational backgrounds and lack basic knowledge on food safety measures. These detriments are amplified in times of high migrations. Migrations are at its highest peak since 2007; around 4.7 million people permanently migrated to OECD countries in 2015 (7% increase from 2014). More importantly, this affects labour market and migrants usually work in occupations that primarily involve routine tasks [22]. In other words, these facts increase the probability of employing a person without basic food safety knowledge and even a person with language barrier. On-board, the cabin crew carries out the service of meals, snacks and drinks, and they are not trained as food handlers (*Author's note: YouTube contains a number of videos shared by flight attendants that prove improper food handling*). Food defence system is as strong as its highest component, people. Besides investing in the education of people who get in contact with the food, innovation can overcome risks related with the food handlers. First example is the use of disposable packages that eliminate the need for tray assembly and even trolley assembly. The second example is the Nestle Sky Tray concept based on the "Hot Pocket" brand, a hand held hot snack, which is packaged on a thermally resistance service tray and delivered directly onto aircraft without any assembly by the caterer [17].

Cyber security issues are of mayor concern, especially from the aspect of loading schedule, staff information, etc. and some preventive IT techniques [18] have been mentioned previously in the text.

16.4 Conclusions

There is no risk that equals zero!

People are the weakest and the strongest loop in the food defence chain. The Food and Drug Administration (FDA) report from 2010 clearly show breaches in the food safety standards in both LSG Sky Chefs and Gate Gourmet [10].

The food on-board should be tested. Samples of the food should be taken just after loading aircraft. Technology today enables fast detection, and even preliminary findings could prevent a potential disaster.

The list of potential threat agents for specific foods/food groups served on-board should be prepared. The process should include factors related to both foods characteristics and passengers perception (physiological mainly), and can be very specific (for each airline, depending on the food list served).

Bioterroristic threat in the airline catering is real. In 2008 LSG Sky Chefs was warned due to botulinum treat [10], and Gate Gourmet had *Shigella* outbreak (47 confirmed, 116 unconfirmed cases) on flights from Hawaii [12].

Education and innovation are extremely important. On one hand, strengthening the knowledge of food handlers in both kitchens and on-board will improve safety [16] while introduction of new concepts can diminish or even eliminate the role of caterer as well as food handling on-board.

References

1. Baş M, Şafak Ersun A, Kıvanç G (2006) The evaluation of food hygiene knowledge, attitudes, and practices of food handlers' in food businesses in Turkey. Food Control 17(4):317–322
2. Batt CA (2016) Food defense. Elsevier Inc
3. Bossi P, Garin D, Guihot A, Gay F, Crance JM, Debord T, Autran B, Bricaire F (2006) Bioterrorism: management of major biological agents. Cell Mol Life Sci 63(1920):2196–2212
4. Brainard J, Hunter PR (2016) Contextual factors among indiscriminate or large attacks on food or water supplies, 1946-2015. Health Secur 14(1):19–28
5. BRC Global Standard (2012) BRC global standard for food safety. http://www.brcglobalstandards.com/Knowledgecentre/FoodSafety-AGlobalView.aspx#.VtWwZ0ZCfPE
6. British Standards Institution (2014) PAS 96:2014 Guide to protecting and defending food and drink deliberate attack. http://www.food.gov.uk/sites/default/files/pas96-2014-food-drink-protection-guide.pdf
7. Dalziel GR (2009) Food defence incidents 1950–2008: a chronology and analysis of incidents involving malicious contamination of the food supply chain. S. Rajaratnam School of International Studies, Nanyang Technological University, Singapore
8. Food and Drug Administration (2002) Bioterrorism act. http://www.fda.gov/RegulatoryInformation/Legislation/ucm148797.htm
9. Food Protection and Defense Institute (2017) Food Adulteration Incidents Registry (FAIR). https://foodprotection.umn.edu/innovations/food-fraudema/incidents-database
10. Food Safety News (by Curran L) (2010) Airline food lacks adequate safety precautions. http://www.foodsafetynews.com/2010/06/airline-food-lacks-adequate-safety-precautions/#.WYEDF-mxVdg

11. FPDI, Food Protection and Defense Institute (2012) Incidents Database: Economically Motivated Adulteration (EMA) and Intentional Adulteration (IA) Incidents Database. FPDI, 2012. Available at: https://foodprotection.umn.edu/innovations
12. Gaynor K, Kanenaka B, Colindres R, Mintz E, Kalluri P, Kitsutani P, Nakata M, Wedel S, Boxrud D, Jennings D, Yoshida H, Tosaka N, Park SY, Effler PV (2004) International outbreak of Shigella sonnei among airline passengers—Honolulu, Hawaii. Presentation abstract, session C, 54th annual epidemic intelligence (EIS) conference, April 11–15, Atlanta, GA
13. Global Terrorism Database (2017) https://www.start.umd.edu/gtd/search/Results.aspx?chart=target&search=food
14. IFSA, International Flight Services Association (2016) World Food Safety Guidelines for Airline Catering, version 4. IFSA, 2016
15. Independent (2015) The 30 best and worst paid jobs in the UK. http://www.independent.co.uk/news/business/news/the-30-best-and-worst-paid-jobs-in-the-uk-a6740731.html
16. Institute of Medicine (2006) Addressing foodborne threats to health – policies, practices, and global coordination. The National Accademies Press, Washington, DC
17. Jones P (2007) Flight-catering. In: Becker H, Grothues U (eds) Catering-Management Portrait einer Wachstumsbranche in Theorie und Praxis, Behr'sVerlag, Hamburg, Chapter 1.4.1, pp 39–55
18. Kanai S, Kakizaki S, Matsutani S, Nakata H, Kaneko S (2015) Case studies of food defense in relation to plant security management. Hitachi Rev 64(4):209–214
19. Khan AS, Swerdlow DL, Juranek DD (2001) Precautions against biological and chemical terrorism detected at food and water supplies. Public Health Rep 116:3–14
20. Lee HK, Halim HA, Thong KL, Chai LC (2017) Assessment of food safety knowledge, attitude, self-reported practices, and microbiological hand hygiene of food handlers. Int J Environ Res Public Health 14(1):55
21. Martins RB, Hogg T, Otero JG (2012) Food handlers' knowledge on food hygiene: the case of a catering company in Portugal. Food Control 23(1):184–190
22. OECD (2017) International migration outlook 2017. OECD Publishing, Paris. https://doi.org/10.1787/migr_outlook-2017-en
23. Puhač Bogadi N, Banović M, Babić I (2016) Food defence system in food industry: perspective of the EU countries. J Verbr Lebensm 11(3):217–226
24. Schmidt U, Struck S, Gruening B, Hossbach J, Jaeger IS, Parol R, Lindequist U, Teuscher E, Preissner R (2009) SuperToxic: a comprehensive database of toxic compounds. Nucleic Acids Res 37:D295–D299
25. Scholthof K-BG (2003) One foot in the furrow: linkages between agriculture, plant pathology, and public health. Annu Rev Public Health 24:153–174
26. Seaman P, Eves A (2010) Perceptions of hygiene training amongst food handlers, managers and training providers–a qualitative study. Food Control 21(7):1037–1041
27. World Bank Data (2017) Air transport, passengers carried. http://data.worldbank.org/indicator/IS.AIR.PSGR
28. World Health Organization (2015) Food safety. WHO, Geneva. http://www.who.int/mediacentre/factsheets/fs399/en/

Chapter 17
Food Safety, Standards and Norms Against Bioterrorism: Food Safety and Hazards

Adela Krivohlavek

Abstract In a world dominated by the trends of globalization, urbanization, global warming and changes in consumer habits the question of the safety and quality of food is inevitable and one of priorities for governments, producers and consumers. Food, in various stages of production, distribution or storage, can be contaminated with different biological, chemical and physical hazards which can cause harmful effects on human health. Among those hazards bacteria, viruses, parasites, toxins and chemical contaminants mostly caused Food-borne diseases. In order to prevent and increase consumer confidence in the safety and quality of food from farm to table, various Norms and Standards are introduced. In order to decide which Norm or Standard is the most appropriate for our special need, we have to be familiar with them.

17.1 Introduction

Globalization, urbanization and changes in consumer habits have increased the number of people buying and eating food prepared in public places, triggered growing consumer demand for a wider variety of foods and globalized, extended and complexed food chain. Because food supply chains now cross multiple national borders good collaboration between governments, producers and consumers helps ensure food safety. On the other side climate changes modify food safety risks associated with food production, storage and distribution. All that makes the global food-supply chain vulnerable to threats from a variety of directions.

Local incidents can quickly evolve into international emergencies due to the speed and range of product distribution. In order to prevent such incidents all over the EU there is one, unique common food safety policy through coherent "**farm-to-table**" measures and adequate monitoring, while ensuring an effective internal market. It was introduced after the BSE bovine spongiform encephalopathy and dioxin crisis at

A. Krivohlavek (✉)
Teaching Institute of Public Health "Dr. Andrija Štampar", Zagreb, Croatia
e-mail: adela.krivohlavek@stampar.hr

V. Radosavljevic et al. (eds.), *Defence Against Bioterrorism*, NATO Science for Peace and Security Series A: Chemistry and Biology,
https://doi.org/10.1007/978-94-024-1263-5_17

the end of nineties. EU food safety policy is consisted out of three elements. (1) Legislation (2) Scientific investigation and (3) Implementation and official control.

The basis of international food legislation are the rules developed by the World Health Organization (WHO), the Food and Agriculture Organization (FAO), the World Trade Organization (WTO), the Codex Alimentarius Commission (CAC) and other international organizations [1, 3, 11, 12]. The General Food Law consists of three parts. The first part sets out the general principles and requirements of food regulations. The general principles applied in line with the integrated "farm to table" approach include risk analysis, precautionary principle, consumer protection and transparency principle. The general requirements of food regulations relate to food safety requirements, traceability and responsibility of food and feed business entities. The second part defines the establishment, role and tasks of the European Food Safety Authority (EFSA) as an independent source of scientific advice and communication on food-related risks. The third part of the General Food Law prescribes different procedures regarding food safety, for example, emergency measures when there are clear indications that food or feed may pose a serious risk to human, animal or environmental health. It also prescribes crisis management in the area of food and feed, and the establishment of the Rapid Alert System for Food and Feed (RASFF) [10].

Whatever the case, Regulation (EC) No 178/2002 [8] sets out traceability as a means of monitoring food and feed throughout the stages of processing and distribution. Precisely traceability has helped and was a key factor in rapid crisis management in the area of food safety when the problem comes. According to the acquis communautaire, traceability in the food chain needs to be ensured. "Traceability" means that food business operators and animal feeders – regardless of whether they are producers, processors or importers – must ensure that all foodstuffs, animals that are bred for food production, animal feed and food ingredients can be followed through the entire food chain, from farm to table.

When crises occur within each of the EU member states, crusades are set up in each of the crisis-affected EU member states and there is also the headquarter for the EU Commission when there is a crisis that poses a direct or indirect risk to human health and which comes from food and feed.

17.2 Food Safety and Hazards Associated with Food

Numerous dangers of different types of origin can enter the food chain and make foods potentially harmful to human consumption and disable free international trade. The concept of food safety implies that food is "without danger" with acceptable risk. Food safety can be endangered by biological, chemical and physical hazards [5] which can cause harmful effects on human health. Understanding these dangers is the foundation for the Hazard Analysis and Critical Control Point (HACCP) system (Table 17.1).

Table 17.1 Types of hazards for food safety [7]

Microbiological hazard	Chemical hazard	Physical hazard
Bacteria (sporulating)	Naturally occurring chemicals	Glass
Clostridium botulinum	Mycotoxins	Wood
Clostridium perfringens	Mushroom toxins	Stone
Bacillus cereus	Histamine	Metal
bacteria (uncontrolled)	Additional chemicals	Plastic
Salmonella spp.	Pesticides	Bones
Shigella	Antibiotics	
Listeria monocytogenes	Growth hormones	
Escherichia coli	Artificial fertilizer	
Campylobacter spp.	Packaging material	
Staphylococcus aureus	Vinyl chloride	
Brucella	Softeners	
Yeast and molds	Adhesives, iron	
Viruses		
Hepatitis A and E	Food supplements, vitamins, minerals	
Rotavirus	Polluters	
Parasites	Detergents, lubricants, dyes	
Trichinella spiralis	Coolant	

17.2.1 Biological Hazards

Biological hazards are living organisms that can contaminate food. Biological risk exists at every stage of production, storage and transport of food. Microbiological hazards are: bacteria, viruses, parasites (protozoa), prions and molds.

- *Salmonella, Campylobacter, and Enterohaemorrhagic Escherichia coli* are among the most common foodborne pathogens that affect millions of people annually – sometimes with severe and fatal outcomes. Symptoms are fever, headache, nausea, vomiting, abdominal pain and diarrhoea. Examples of foods involved in outbreaks of salmonellosis are eggs, poultry and other products of animal origin. Foodborne cases with *Campylobacter* are mainly caused by raw milk, raw or undercooked poultry and drinking water. *Enterohaemorrhagic Escherichia coli* is associated with unpasteurized milk, undercooked meat and fresh fruits and vegetables.
- *Listeria* infection leads to unplanned abortions in pregnant women or death of newborn babies. Although disease occurrence is relatively low, listeria's severe and sometimes fatal health consequences, particularly among infants, children and the elderly, count them among the most serious foodborne infections. *Listeria*

is found in unpasteurised dairy products and various ready-to-eat foods and can grow at refrigeration temperatures.

- *Vibrio cholerae* infects people through contaminated water or food. Symptoms include abdominal pain, vomiting and profuse watery diarrhoea, which may lead to severe dehydration and possibly death. Rice, vegetables, millet gruel and various types of seafood have been implicated in cholera outbreaks.

Antimicrobials, such as antibiotics, are essential to treat infections caused by bacteria. However, their overuse and misuse in veterinary and human medicine has been linked to the emergence and spread of resistant bacteria, rendering the treatment of infectious diseases ineffective in animals and humans. Resistant bacteria enter the food chain through the animals (e.g. *Salmonella* through chickens). Antimicrobial resistance is one of the main threats to modern medicine.

17.2.1.1 Viruses

Norovirus infections are characterized by nausea, explosive vomiting, watery diarrhoea and abdominal pain. Hepatitis A virus can cause long-lasting liver disease and spreads typically through raw or undercooked seafood or contaminated raw produce. Infected food handlers are often the source of food contamination.

17.2.1.2 Parasites

Some parasites, such as fish-borne trematodes, are only transmitted through food. Others, for example tapeworms like *Echinococcus* spp., or *Taenia solium*, may infect people through food or direct contact with animals. Other parasites, such as *Ascaris*, *Cryptosporidium*, *Entamoeba histolytica or Giardia*, enter the food chain via water or soil and can contaminate fresh produce.

17.2.1.3 Prions

Prions are infectious agents composed of protein, are unique in that they are associated with specific forms of neurodegenerative disease. *Bovine spongiform encephalopathy* (BSE, or "mad cow disease") is a prion disease in cattle, associated with the variant Creutzfeldt-Jakob Disease (vCJD) in humans. Consuming bovine products containing specified risk material, e.g. brain tissue, is the most likely route of transmission of the prion agent to humans.

17.2.2 Chemical Hazards

Chemical hazards can occur in food during primary production (pesticides and remedies), processing (additives, toxic nus products of production processes – contaminants), packaging (migration of packaging materials) and contact with environmental substances (contaminants). Absorption of toxic substances from food and toxic effects of these substances is influenced by numerous factors. Among the most important factors are concentrations or doses of toxic substances, exposure/retention length, physico – chemical properties, route of entry, individual resistance, etc. Chemical contamination can lead to acute poisoning or long-term diseases, such as cancer. Foodborne diseases may lead to long-lasting disability and death. Examples of unsafe food include uncooked foods of animal origin, fruits and vegetables contaminated with faeces, and raw shellfish containing marine biotoxins.

Of most concern for health are naturally occurring toxins and environmental pollutants.

- Naturally occurring toxins include mycotoxins, marine biotoxins, cyanogenic glycosides and toxins occurring in poisonous mushrooms. Staple foods like corn or cereals can contain high levels of mycotoxins, such as aflatoxin and ochratoxin, produced by mould on grain. A long-term exposure can affect the immune system and normal development, or cause cancer.
- Persistent organic pollutants (POPs) are compounds that accumulate in the environment and human body. Known examples are dioxins and polychlorinated biphenyls (PCBs), which are unwanted by-products of industrial processes and waste incineration. They are found worldwide in the environment and accumulate in animal food chains. Dioxins are highly toxic and can cause reproductive and developmental problems, damage the immune system, interfere with hormones and cause cancer.
- Heavy metals such as lead, cadmium and mercury cause neurological and kidney damage. Contamination by heavy metal in food occurs mainly through environmental pollution of air, water and soil.

17.2.2.1 Contaminants Generated During Processing or Storage of Food

Food additives are added to food to improve their technological performance and maintain their sensory properties. Additives are added to the food during production, preparation, processing, shaping, packaging, transportation and storage. Modern food production can not be imagined without the addition of additives under well-defined conditions with a well-established reason. The use of additives is directly related to their basic functional, technological properties, so they are now divided into categories: dyes, preservatives, antioxidants, emulsifiers, stabilizers, thickeners, gelling agents, acidity regulators, acids, substances to prevent stiffening, taste enhancers, sweeteners, modified starches, polishing agents, moisture retention agents, flour retardants, fasteners, volume enhancers, thrusters, emulsifying salts, foaming agents, gases for packing and sequestering.

17.2.2.2 Contaminants from Materials and Articles in Contact with Food

The materials and objects for direct contact with food are made up of anthropogenic or natural organic macromolecular substances and inorganic materials. Food quality and safety can be threatened from unwanted migration from food packaging. The transfer of the total amount of substance from food packaging is called global migration. Chemical migration is a diffusion process, run under kinetic and thermodynamic control, and is described by a mathematical model of diffusion from Fick's law. Generally, any kind of material or objects that come into direct contact with food can be a source of chemical migration. Specific migration is the migration of identified toxic substances. Specific migration is legally covered by about 400 contaminants (metals, monomers, additives, etc.) for different types of materials, while it is believed that there are more than 3000 potential substances that can affect food health.

17.2.3 Physical Hazards

Physical hazards pose foreign bodies in food that can cause injuries, illness and psychological trauma. Even 25% of consumers' complaints refer to physical contamination. Mechanical hazards are divided into those minerals (earth, stones, dust, glass, metal, paint), herbal (weed, leaves, stems) and animal origin (insects, rodents, worms) [5].

17.3 Foodborne Diseases

For the last decades, the occurrence of pathogenic microorganisms and their harmful metabolites in food has become frequent due to free trade, new technologies and new consumer interests around the world. In the last decade, the phases connecting public health with food production are being developed and included in policies of WHO and FAO.

Food can be contaminated in different stages of production, storage and transport. Figure 17.1 shows the main routes of transmission of foodborne diseases [4].

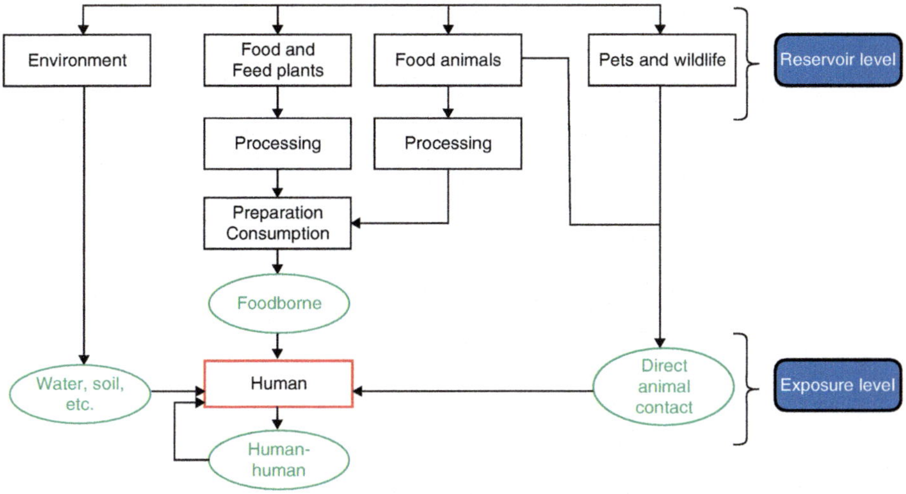

Fig. 17.1 The main routes of transmission of foodborne diseases

Serious foodborne disease outbreaks have occurred on every continent in the past decade, often amplified by globalized trade. According to the research, it is estimated that in 2010 the disease foodborne affected 582 million people, of whom 38% are children under 5 years of age. The most common causes are *norovirus, Campylobacter* spp., *ETEC, NTS and Shigella* spp., followed by *V. cholerae* and *Salmonella* [6].

According to the latest WHO data, globally, nearly one in 10 people are ill with consumption of contaminated food. Food-borne diseases are estimated to be mostly caused by 31 agents: bacteria, viruses, parasites, toxins and chemical contaminants. Every year, as many as 600 million people worldwide suffer after eating contaminated food. Out of that number, 420,000 people die, including 125,000 children under the age of five. The most endangered areas are Africa and the southeastern region of Asia.

17.4 International Standards and Norms

Food manufacturers and distributors are increasingly demanding the application of certain standards, above all by large chain stores, but also as consumers' demands for the safety and quality of food products become a priority. Given organized and development-oriented large trading systems, they increasingly take over production control, safety and quality products they sell, systematically explore and monitor customer needs, control production, direct product development, and increasingly impact on shaping nutrition habits and economic trends in the environment they operate.

Food security in the market economy with the mandatory regulatory framework consists of additional voluntary instruments such as good practice guides and various international standards and standards [9]:

1. Good and Practice Guidelines at National and International Levels or prescribed by the Codex Alimentarius are guidelines for subjects from the food sector. There are nine good practices in the field of food safety today:
   - Good Farming Practice (GAP),
   - Good Manufacturing Practice (GMP),
   - Good Laboratory Practice (GLP),
   - Good Hygiene Practice (GHP),
   - Good Distribution Practice (GDP),
   - Good Storage Practice (GSP),
   - Good Trade Practice (GTP),
   - Good Catering Practice (GCP)
   - Good Housekeeping Practice (GHKP).
2. Norms in the field of the agro-food industry most commonly used are:
   - GLOBALGAP – standards for voluntary certification of agricultural products
   - International Food Standard (IFS) – a standard established for the control of large-scale branded goods manufacturers, and has developed German, French and Italian chain stores.
   - The British Retail Consortium (BRC) [2] – a British Retail Consortium (BRC) standard designed to help traders fulfill their legal obligations regarding consumer protection, providing a common basis for verifying all manufacturers of their brands.
   - A series of ISO 22000 norms – created as an expression of the industry's aspiration to create an international standard that would be acceptable and recognizable in all countries and which would replace a number of national standards. ISO Standardization Organization in 2005, ISO issued a standard and a food safety management system (ISO 22000: 2005). The standard includes the requirements of a food safety management system and key food safety elements along the chain from manufacturers to consumers, reciprocal communication, prerequisite program management system, and HACCP principles. Trade associations and suppliers did not accept the norm ISO 22000: 2005 because the standard did not sufficiently define the requirements of the prerequisite program. Therefore, in 2009 the same organization issues the ISO 22002–1: 2009 Prerequisite Food Safety Program (Food Production). In this way, the Food Safety System Certification (FSSC) was created, through which ISO 22000: 2005 standard was accepted in practice. Approved only as FSSC 22000.
   - Global Food Safety Initiative (GFSI) – A Global Food Safety Initiative, founded in 2000, has been created by leading retail chains and suppliers worldwide (Carrefor, Tesco, ICA, Metro, Migros, Ahold, Wal-Mart, Delhaize and

others.). The GFSI mission is to continuously improve the food safety system to ensure consumer confidence in delivering safe food. GFSI supports recognized international certification schemes from IFS, BRC and FSSC 22000.

Effective implementation of the control program is a guarantee to the manufacturer to deliver to its customers only safe and high quality products that operate within the framework of standard food quality assurance systems (BRCs, International Food Safety Initiatives, etc.). Regardless of the legal obligation to introduce procedures based on the principles of the HACCP system, within the aforesaid systems already have integrated GHP/GMP requirements.

Another problem is the number of standards. Due to the lack of a unique international certification program for food safety, a number of certification programs have emerged. Though they vary widely between objectives and scope, they are commonly related to good practice programs and food safety management systems, mostly referring to the Codex Alimentarius standard. The second common feature is their geographical constraint, i.e. the bondage to the trading companies of certain countries. The increase in the number of such independent standards with somewhat different demands has led to rising consumption of time, energy and financial resources for food manufacturers, especially those supplying more than one country or multiple chain stores in one country. Many found themselves in a situation where two or more programs had to be certified, and two or more programs each year, and two or more visits per year to third party auditors, basically the same requirements, only set in somewhat different (sometimes mutually incompatible) ways.

17.5 Conclusion

The protection of human, animal and plant health at every stage of the food production process is a key priority of public health and the economy.

There are a number of hazards in the modern global market so standards are needed to prevent the spread of animal and plant diseases, prevent spreading of animal diseases to humans, protect consumers and prevent unfair competition, protect animal welfare, ensure clear and unambiguous consumer information.

The process of standardization and certification based on internationally recognized standards from year to year gains significance, and certification is recognized by both manufacturers and retail chains as well as by consumers as an additional guarantee that the food that is placed on the market is safe and of good quality.

Unsafe food poses global health threats, endangering everyone. Infants, young children, pregnant women, the elderly and those with an underlying illness are particularly vulnerable.

Integrated health protection is essential through systematic control of the entire "agri-food chain" – in every part of the process of food production, from breeding to consumption, prevention of contamination and promotion of food hygiene, food information, plant health and animal health and wellbeing.

References

1. Babić I, Đugum J i sur (2014) Uvod u sigurnost hrane. ISBN 978-961-92846-2-9
2. BRC Global Standard (2012) BRC global standard for food safety. http://www.brcglobalstandards.com/Knowledgecentre/FoodSafety-AGlobalView.aspx#.VtWwZ0ZCfPE
3. Codex Alimentarius. General principles of food hygiene. FAO/WHO
4. Hald T, Aspinall W, Devleesschauwer B, Cooke R, Corrigan T et al (2016) World Health Organization estimates of the relative contributions of food to the burden of disease due to selected foodborne hazards: a structured expert elicitation. PLoS One 11(1):e0145839
5. Havranek J, Tudor Kalit M i sur (2014) Sigurnost hrane od polja do stola M.E.P. pp 22–48. ISBN: 978-953-6807-31-4
6. Kirk MD, Pires SM, Black RE, Caipo M, Crump JA, Devleesschauwer B et al (2015) World Health Organization estimates of the global and regional disease burden of 22 foodborne bacterial, protozoal, and viral diseases, 2010: a data synthesis. PLoS Med 12(12):e1001940
7. Markov K, Frece J, Čvek D, Lovrić N, Delaš F (2010) Aflatoksin u sirovom mlijeku i vezanje aflatoksina pomoću bakterija mliječne kiseline. Mljekarstvo 60(4):244–251
8. Regulation (EC). No 178/2002 of the European Parliament and of the Council of 28 January 2002 laying down the general principles and requirements of food law. http://eur-lex.europa.eu/legal-content/EN/ALL/?uri=CELEX:32002R0178
9. Trienekens J, Zuurbier P (2008) Quality and safety standards in the food industry, developments and challenges. Int J Prod Econ 113(1):107–122
10. Van Asselt ED, Meuwissen MPM, Van AMAPM (2010) Selection of critical factors for identifyingemerging food safety risk in dynamic food production chains. Food Control 21:919–926
11. World Health Organization (2015) Food safety. WHO, Geneva. http://www.who.int/mediacentre/factsheets/fs399/en/
12. Zakon o hrani. (Narodne novine broj 81/13, 14/14, 30/15)

Chapter 18
Environment and Bio-Terrorism

Stéphanie Watier-Grillot, Olivier Cabre, Gabriel Bédubourg, Jean-Paul Demoncheaux, Christian Hupin, and Benjamin Queyriaux

Abstract Environment is a dynamic system around humans, embracing all kind of living bodies (animal, vegetal, fungi and parasites, bacteria, viruses…) and natural elements as air, earth and water. Bio-terrorism interacts constantly with environment, in order to achieve its main goal: to put terror on humans using biological agents. Bio-terrorism deals with the environment as a source of bio-terrorism agents, as a means of bio-terrorism and as a target. These three interfaces bio-terrorism/environment lead to common possibilities to prevent bio-terrorism and protect environment, mainly within the one-health concept.

Environment can be defined as the natural world of land, sea, plants and animals. Environment is a dynamic system around humans, embracing all kind of living bodies (animal, vegetal, fungi and parasites, bacteria, viruses…) and natural elements as air, earth and water. Consequently, it is obvious that bio-terrorism is dealing constantly with environment, in order to achieve its main goal: to put terror on humans using biological agents.

This article presents three different ways in which bio-terrorism is dealing directly with environment: environment as a source of bio-terrorism agents, environment as a means of bio-terrorism and environment as a target of bio-terrorism. These three interfaces bio-terrorism/environment lead to common possibilities to prevent bio-terrorism and protect environment.

S. Watier-Grillot (✉) · G. Bédubourg · C. Hupin
French Military Centre for Epidemiology and Public Health, Marseille, France
e-mail: s.watier.grillot@gmail.com

O. Cabre · J.-P. Demoncheaux
French Surgeon General Office, Paris, France

B. Queyriaux
North Atlantic Treaty Organisation International Military Staff, Brussels, Belgium
e-mail: benjamin.queyriaux@gmail.com

V. Radosavljevic et al. (eds.), *Defence Against Bioterrorism*, NATO Science for Peace and Security Series A: Chemistry and Biology,
https://doi.org/10.1007/978-94-024-1263-5_18

18.1 Environment As a Source of Bio-Terrorism Agents

Despite the fact that environment is the natural world of bio-terrorism agents, mainly bio-toxins, viruses and bacteria, the recent history has shown that terrorist groups are planning to use preferentially bio-agents available in bio-laboratories. Nevertheless, these laboratories are under strict state surveillance, and possibilities to illegally extract bio-samples usable for terrorism purposes are minimal. That is why the extraction of bio-terrorism agents from the environment is a possibility that must be taken into account [1].

Most of human infectious diseases are of animal origin (zoonosis). For many emerging pathogens, wildlife and sometimes even domestic animals show no signs of infection and play the role of asymptomatic reservoirs (e.g. Ebola virus, Crimean-Congo Haemorrhagic Fever…). Today, OIE-World Organisation for Animal Health estimates that 60% of existing human infectious diseases are zoonotic; at least 75% of emerging infectious diseases of humans (including Ebola, HIV and Influenza) have an animal origin; five new human diseases appear every year and three of them are of animal origin; 80% of agents with potential bio-terrorist use are zoonotic pathogens [2].

The Center for Disease Control and Prevention (CDC) has published a list of the most probable bio-terrorism agents that could be used [3]. All of them are available in the environment: the majority are zoonosis or have an animal reservoir.

An example of biological agent that could be extracted from the environment is given by the permafrost (an area of land that is permanently frozen below the surface) located in Siberia, Canada and Alaska. Due to global warming some infected bodies will emerge from the Permafrost, and intact biological agents will re-emerge. An anthrax outbreak was described in 2016 in Siberia, after no outbreaks for 75 years. An exceptionally warm summer thawed out a reindeer carcass buried in a permafrost pit that was a source of spores [4]. As human bodies infected by smallpox virus are frozen in the permafrost [5], the possibility of a re-emergence of this virus consequently of a permafrost thaw, and then its use for bio-terrorism purpose, is a valid assumption made by several medical intelligence departments.

Some plants are known to naturally produce and contain highly toxic compounds (toxins, cyanogenic molecules, cardiotoxic alkaloids…). When ingested or inoculated, small amounts of these plants have the potential to cause severe damage to health of human beings or animals even up to death, including cardiotoxicity, neurotoxicity, cytotoxicity and metabolic disorders (inhibition of cell division). For example, Ricin can be extracted from Castor Bean. Al Qaeda in Yemen tried that in 2011, without success [6, 7]. It shows however the interest those terrorists bring to the possibility to extract bio-agent from the environment.

18.2 Environment As a Mean of Bio-Terrorism

In the same way it supplies disease agents, environment can naturally convey biological pathogens by several routes including inhalation, inoculation or ingestion.

Animals can also act as spreaders of biothreat agents, particularly migratory species (birds) and insect vectors (mosquitoes, ticks, fleas).

Nevertheless, as biological agents are living microorganisms, they are fragile and may be killed by the forces needed to disseminate them or the exposure to environmental factors as drying, temperature (heat, cooking), UV radiations, etc.

Meteorology, dilution effect (water), density and solubility are additional factors limiting bioweapon uses.

Urban environment can also promote peculiar scenarios. For example, the introduction of hazardous animals in crowded areas and/or closed spaced could generate fear, panic and a significant media impact. It's another potential aspect (based on macroscopic component) of bio-threat from animal origin. Another illustration of the use of animals as bio-terrorism vectors is the attack performed by Talibans in Afghanistan in 2009, using a donkey as explosive vector: the donkey transported an Improvised Explosive Device (IED) to the coalition's facility [8].

18.3 Environment As a Target of Bio-Terrorism

Agro-terrorism can be defined as the deliberate introduction of a disease agent, either against livestock or into the food chain for the purposes of undermining socio-economic stability and/or generating fear [9, 10].

In developed countries, due to the vertical integration of many food industries, the food chain is highly vulnerable to bio-terrorism attacks. Deliberate introduction of a pathogenic agent is possible anywhere along the production line that runs from farm to fork. Pre-harvest threats target livestock and crops, carrying the risk of economic devastation compounded by direct costs (international trade restrictions, slaughter of animals, loss of production) and indirect costs throughout related communities (e.g. tourism). Post-harvest threats affect the food industries (processing, transportation, delivery) and public health (possible human illness and death). Detrimental social, political, diplomatic and even military consequences could follow an agro-terrorist attack.

Literature review indicates that the use of biological agents against livestock and poultry populations has been rare when compared to other targets. Between 1900 and 1998, only three instances of the use of biological agents against agriculture targets were found [11]. Nevertheless, agro-terrorism remains an attractive option [12, 13]:

1. animal or plant biological pathogens are easy to acquire. Due to the endemic nature of many of these disease agents in large geographic regions worldwide,

samples of agents are readily available from clinical specimens collected from the field during natural outbreaks of disease.
2. In general these agents can easily be disseminated. Animals and plants provide the primary means of transmission. Sophisticated weaponisation is not required.
3. Many animal or plant diseases are not zoonotic. Therefore, they are not harmful for the perpetrator and there are no requirements for elaborating personal protective equipment and containment procedures.
4. Nevertheless, if the objective is human casualties, some zoonotic diseases offer unique capabilities and the food chain offers a low-tech but highly conductive mechanism for disseminating a wide range of pathogens, including bacteria, viruses and toxins.
5. Many animal and plant pathogens are predictable, with an expected clinical pattern and are attributable to a natural outbreak, ensuring plausible deniability if desired.
6. There is a large selection of animal and plant biological agents from which to choose. Various selection criteria for the most dangerous anti-livestock and anti-poultry biological agents have been suggested [14]. According to the old classification of notifiable diseases to the OIE, «list A» includes not less than 12 pathogens identified as having the potential to seriously impact animal health and/or trade (e.g. foot-and-mouth disease, African or classical swine fever, highly pathogenic avian influenza, lumpy skin disease, Rift Valley fever…). Notable among these are zoonotic and epizootic diseases of both wild and domesticated animals such as rabies, anthrax, plague, tularemia, glanders, melioidosis, foot-and-mouth disease, rinderpest, psittacosis, canine distemper, and Venezuelan equine encephalomyelitis. For example, more than 60 wildlife species are known to be susceptible to foot-and-mouth disease, but the total number of susceptible wildlife species around the globe is probably much higher. The extinction of local populations of North American black footed-ferret (*Mustela nigripes*) and African wild dog (*Lycaon pictus*) by canine distemper, a common viral disease of domestic dogs, are other excellent examples. Regulated crop pests are also a major concern.

Bio-weapons and emerging disease outbreaks could result in severe erosion of genetic diversity in populations of wild and domestic animals and plants, leading to the extinction of endangered species. The threat lies in the release and proliferation of a broad spectrum of diseases of domesticated livestock and crops among naive, susceptible populations of wildlife and plants.

18.4 Environment As a Contributing Factor to Bio-Threat Reduction

Key-factors for bio-threat management and reduction are: prevention, reaction, mitigation and recovery. Being prepared for the norm, one is prepared for the abnormal. The methods for emergency responses are common for both circumstances, natural and intentional, and must be dual purposes [15, 16].

Effective emergency response requires a multidisciplinary approach. Interagency coordination, both national and international, of the food, heath and agriculture sectors, as well as intelligence and security services is essential [17].

Environment can provide useful signals and indicators for early warning and health monitoring. Most of the diseases that could potentially be used in bioterrorism attacks are common to humans and animals. Therefore, they can be used as biosensors/sentinels to help to identify a bioterrorism-related threat to human public health [18, 19]. They could provide early warning of an attack or could function as markers for the risk of on-going exposure. As examples of animal species that could potentially be used as sentinels of bio-threats, birds for West Nile Virus and other viral encephalitis agents, pets for anthrax and plague, cattle for anthrax and Rift Valley Fever can be quoted. Beside the bio risk, animals are also susceptible to the four main types of chemical agents: vesicular, respiratory, neurotoxic and hematotoxic agents, and thus serve as sensors for chemical risks.

The Global Early Warning and Response System (GLEWS) is a great example of an integrated health monitoring system. It combines and coordinates the alert and response mechanisms of OIE- World Organisation for Animal Health-, Food and Agriculture Organisation and World Health Organisation. It assists in prediction, prevention and control of animal disease threats, including zoonosis.

Medical (or Health, or Epidemic) Intelligence, civilian or military, monitors continuously the epidemiological status worldwide for humanity and environment. Using almost exclusively open-source data, medical intelligence aims to detect as soon as possible any kind of bio-event that may impact a population, human or animal, or agriculture. Nevertheless, these medical intelligence systems are mainly implemented in western countries by different state departments (defence, health, agriculture…) according to their objectives, and they are rarely connected. They provided accurate pictures of bio-risk and bio-events within their domains.

Some unusual or unexpected events in the environment can be very relevant signals of an outbreak, natural or intentional. Environment analysis cannot be separated from the human or the animal epidemiological status. Consequently, unified medical intelligence systems, gathering intelligence about humans and environment at once, could be an improvement to achieve bio-terrorism prevention and early detection of unusual events.

Regarding outbreak response, from natural or intentional origin, the deployable outbreak investigation and control teams, implemented by military and civilian authorities, are ideally composed of medical, veterinarian, entomological and environmental expertise. These skills and the knowledge allows these teams to face any kind of bio-event, impacting environment and/or humans.

These examples of bio-threat reduction have a common point: human health, animal health and environment are considered as a whole and are treated together. This is described by the One Health concept [20], which intends to build bridges between these domains, in order to consider health as a whole, and not only focused on humans, or animals, or environment.

In conclusion, it is obvious that the environment is a key dimension in the understanding of bio-terrorism. Its prevention cannot only be focused on humans (as terrorists and targets) but must embrace all dimensions of this challenge. The One-Health concept can be applied here with success. One Health recognizes that the health of people is connected to the health of animals and the environment. The goal of One Health is to encourage the collaborative efforts of multiple disciplines-working locally, nationally, and globally-to achieve the best health for people, animals, and our environment. It is definitively a right approach to address the bio-terrorism threat.

References

1. Ackerman GA (2006) It is hard to predict the future: the evolving nature of threats and vulnerabilities. Rev Sci Tech Off Int Epiz 25(1):353–360
2. OIE-World Organisation for Animal Health- http://www.oie.int/en/for-the-media/onehealth/
3. Centers for Disease Control and Prevention (CDC) – https://emergency.cdc.gov/agent/agentlist-category.asp
4. Gainer R (2016) Yamal and anthrax. Can Vet J 57(9):985–987
5. Biagini P, Thèves C et al (2012) Variola virus in a 300-year-old Siberian mummy. N Engl J Med 367:2057–2059
6. http://www.nytimes.com/2011/08/13/world/middleeast/13terror.html
7. Schep LJ, Temple WA, Butt GA, Beasley MD (2009) Ricin as a weapon of mass terror-separating fact from fiction. Environ Int 35(8):1267–1271
8. http://www.telegraph.co.uk/news/uknews/defence/6716631/Taliban-use-donkeys-to-launch-bomb-attacks-on-British-troops.html
9. Monke J. Agroterrorism: threats and preparedness. US Congressional Research Service report for US Congress, 2007, 12 March, Order Code RL32521, 63 p
10. Dudley JP, Woodford MH (2002) Bioweapons, bioterrorism and biodiversity: potential impatcs of biological weapons attacks on agricultural and biological diversity. Rev Sci Tech Off Int Epiz 21(1):125–137
11. Hugh-Jones M, Brown CC (2006) Accidental and intentional animal disease outbreaks: assessing the risk and preparing an effective response. Rev Sci Off Int Epiz 25(1):21–33
12. Chomel BB, Sun B (2010) Bioterrorism and invasive species. Rev Sci Tech Off Int Epiz 29(2):193–199
13. Blancou J, Pearson JE (2003) Bioterrorism and infectious animal diseases. Comp Immun Microbiol Infect Dis 26:431–443
14. Davis RG (2004) The ABCs of bioterrorism for veterinarians, focusing on category A agents. JAVMA 224:1084
15. Hugh-Jones M (2006) Distinguishing between natural and unnatural outbreaks of animal diseases. Rev Sci Tech Off Int Epiz 25(1):173–186
16. McEwen SA, Wilson TM, Ashford DA, Heegaard ED, Kuiken T, Kournikakis B (2006) Microbial forensics for natural and intentional incidents of infectious disease involving animals. Rev Sci Tech Off Int Epiz 25(1):329–339
17. OIE. OIE Fact Sheet on Biological Threat reduction. 2015. [Online]. http://www.oie.int/en/our-scientific-expertise/biological-threat-reduction/
18. Rabinowitz P, Gordon Z, Chudnov D, Wilcox M, Odofin L, Liu A, Dein J (2006) Animals as sentinels of bioterrorism agents. Emerg Infect Dis 12(4):647–652
19. Enriquez B, Dropsy H (2013) Bioterrorisme par agents bacteriens et viraux: prevention du risque pour l'homme par utilisation d'animaux sentinelles. Environ Risque. Sante 12:490–500. (in French)
20. Gyles C (2016) One medicine, one health, one world. Can Vet J 57:345–346

Chapter 19
Ethical Aspects of Bioterrorism and Biodefence

Elizabeta Ristanovic

Abstract Bioterrorism is the phenomenon as old as civilization. The use of biological agents in war and subversive actions was recognized as an unhonourable weapon and a crime against humankind even in the ancient time. The development of bioweapons was always the expression of scientific and technological progress resulting in new military aspirations so during the Cold War period the world was even at the edge of the real biological warfare. After the sign of BWC the most powerful state actors prohibited the use of bioweapons but all of them continued the investigations with *"biodefence purposes"*. Today in the post 9/11 world bioterrorism is recognized as one of the leading security threat of the modern world that should be considered from the socio-political, economic, security, scientific, public-health, ecological and ethical points of view. The scientific progress, especially in the fields of molecular biology, genetic engineering, biotechnology and nanotechnology opened the questions about their possible misuse for improvement of biological weapons and making them more specific and effective (*dual-use dilemma*). In order to protect humankind and prevent possible misuse of biotechnology and bioterrorism in the world it is necessary to establish an international consensus in bioethical approaches. The ethics questions and considerations in bioterrorism and biodefense must cover multidisciplinary issues including the ethical principles of medicine, fundamental sciences, technology, law, politics, international relations, security, public health, environment, economy and war conducting, each with the unique ethical framework. But it also opens many controversies that will be discussed in the paper. The act of bioterrorism itself also change ethical approaches and make new frame of the ratio between personal, national and international security and open many other questions that should be analyzed.

Keywords Bioterrorism · Biological weapons · Biodefence · Ethical issues · Research

---

E. Ristanovic (✉)
Military Medical Academy, University of Defence, Belgrade, Serbia
e-mail: elizabeta.ristanovic@vma.mod.gov.rs

V. Radosavljevic et al. (eds.), *Defence Against Bioterrorism*, NATO Science for Peace and Security Series A: Chemistry and Biology,
https://doi.org/10.1007/978-94-024-1263-5_19

19.1 Introduction

Bioterrorism as an intentional misuse of microorganisms (bacteria, viruses, fungi, parasites) or their toxins in terrorist purposes is actual and serious threat in the modern world of global contradictions. Microorganisms can be used by terrorist groups that are not under control or by some state actors for achievement of tactical and strategic goals. The science progress especially in the fields of genetics, biotechnology and nanotechnology can be also misused in this purpose. Although the dangerous implications of contemporary biology had been recognized much earlier [1], the anthrax attacks in the USA in 2001 raised awareness and heightened the real concern about them. Microorganisms today can be also considered as a great security threat not only due to bioterrorism but also regarding great naturally-occurring epidemics and pandemics such as smallpox, plague, Spanish flu that changed human history in the past. Climate changes also influence the distribution of microorganisms and their vectors enabling the spreading of microbes to new areas. Great migrations could also pose a problem because microorganisms also move with people. Agricultural bioterrorism, targeting livestock and crop resources with agents such as foot and mouth disease, vesicular stomatitis, rinderpest, lumpy skin disease, blue tongue, sheep and goat pox, African horse sickness, African swine fever, fowl plague, and many others is also an important possible weapon, influencing both the economies and politics of nations.

That is why it is clear that biodefence preparedness and infective agents mitigation measures must be important step in the national strategies as well as the subject of intensive international cooperation and efforts. It is also a moral obligation of the states and nations as well as international community to mitigate that risk. The basic steps in biodefense strategy include epidemiological surveillance as well as development and permanent improvement of detection and identification methods, as well as vaccines and drugs, biosafety and biosecurity standards [2]. Concerning the zoonotic nature of many potential biological agents (tularemia, anthrax, plague, brucellosis, ricketsioses, Ebola etc.) as well as the real recognized threat of agroterrorism it is important to have veterinary medicine permanently involved in the preparations to respond to bioterrorism [3, 4]. The engagement of other experts in biomedical field is also important.

Successful risk prediction and strategic plans to avoid panic and organize an efficient response and effective countermeasures in the event of bioterrorist attack as well as defining and prediction of political and communication channels for international cooperation in bioterrorism countering and responding must be an integral part of biodefence strategy [5]. This multidisciplinary task must be realised by coordinated efforts of the governments and military services, scientists, medical sector, industry.

Although ethical considerations appeared in the past times in connection to use of bioweapons in wars and terrorist actions nowadays it is clear that they are fully incompatible because the missing of ethics makes a terrorist a terrorist. But bioterrorism anyway raises some ethical questions for those who would respond, at least if they do not want to be dragged down to the level of the terrorists. The ethic questions and considerations in biodefense must capture multidisciplinary issues including the ethical principles of human and veterinary medicine and biomedical research, other fundamental sciences, technology, law, politics, international relations, security, public health, environment, economy and war conducting, each with the unique ethical framework. Taking all mentioned into consideration the question arises whether it is actually necessary to define specific ethical guidelines for biological defense and whether it would be possible to apply the same in the case of a real bioterrorist act. The issues of resources and medical and other official personnel allocation, triage assessment related to treating patients [6], **testing of potential therapies or vaccines, awareness or researchers, dual –use dilemmas considering publications and the use of technologies, personnel freedom in the case of quarantine and other situations as well as education of researchers and health care providers and community** to avoid contributing to the advancement of biowarfare and bioterrorism [7].

19.2 Bioterrorism in the Ethical Frame: Historical Overview

The misuse of infective agents in war or terrorist actions is a threat as old as the human society and civilization and it was reported even in the fourteenth century BC when the Hittites used rams infected with tularemia to weaken their enemies [8]. In the fourth century BC, the Scythians infected the arrows by dipping them in a decomposing cadavers and human blood [9]. The Hannibal set fire to the enemy's fleet with pots full of venomous snakes [10]. But even the ancient Romans considered this form of warfare dishonorable. Wage wars by weapons, not with the toxins, it was pointed out in the Roman Senate, while French Emperor Louis 14 paid money to the inventors of biological weapons that do not advertise their innovations. In the Middle Ages, a bubonic plague which killed >25 million European people was transmitted by the Mongols who threw diseased cadavers with catapults into the besieged city of Caffa, and by ships transporting Genovese soldiers, fleas and rats were brought to the Mediterranean ports [11]. During the subsequent centuries, smallpox represented the most effective biological weapon used by European conquerors against native Americans [12].

The modern era of BW started with the foundation of microbiology in the nineteenth century. The World War I was the best known period of chemical warfare agents use while trials with biological agents became more sophisticated in that time. The involved nations, especially Germany and less France, developed secret BW programmes using some zoonotic agents such as *Bacillus antrhacis* and *Burkholderia mallei* [13]. The attempts of Germans to ship horses and cattle inoculated with etiological agents of anthrax and glanders to US and elsewhere were recorded as well as the subsequent attempts to spread cholera in Italy and plague in Russia. The Spanish flu pandemic killed 20–40 millions of people immediately after the war [2].

Taking into consideration all possible consequences and the applied combination of biological and chemical warfare it is clear that weapons of mass destruction (WMD) in that period became a major political concern at the international level [14].

The Geneva Protocol for the Prohibition of the Use in War of Asphyxiating, Poisonous or Other Gases, and of Bacteriological Methods of Warfare was adopted in 1925 and prohibited the use of chemical and biological weapons as undue to modern civilization. It was the first multilateral agreement, that extended prohibition of chemical agents also to biological agents, but in spite of it, offensive biological programs were further carried out. The Protocol did not include either the researches in the field of BWs or their production so the leading world countries (France, the UK, Soviet Union, Canada, Poland, Italy) continued or even started BW production and the Geneva Protocol remained only a dead letter. The other countries such as USA did not ratify the document until 1975.

The Soviet Union also started development of serious BW program in 1929. BW agents (*Shigella, Bacillus anthracis, Vibrio cholerae*) were recovered from Russian spies by the Japanese. During the war were reported suspected use of tularemia agent at Stalingrad in 1942 with 100.000 infected German soldiers. Russian soldiers and civilians were affected. The outbreak of Q-fever among Germans on their vacation at Crimea in 1943 was also supposed to be the use of the biological agent. The other countries also showed interest in this field. The UK began to study anthrax as BW before the WWII and leading research center on B-warfare was established in Porton Down. The anthrax was tested on sheep at Gruinard island in 1942, and the island was contaminated with spores even 42 years later. The intensive research during the war was conducted with possible BW agents use in V-1 missiles. The Japanese government started BW program during the interwar period by setting the Army Research Laboratory known as The Unit 731 which conducted terrible experiments over the Chinese and Mongolian prisoners using biological attacks with cholera, smallpox, botulism, bubonic plague, anthrax, tularaemia, and venereal diseases, and developed various methods of BW dissemination and spreading [15]. They were accused for using of biological agents against the Soviet Union and Mongolia, by releasing the 22.5 kg of bacteria into river in 1939, while in the period 1940–1944 they used biological agents several times against Chinese civilians (dropping of contaminated grain and fleas into cities resulting in outbreaks of bubonic plague; cholera "vaccination" with 10,000 civilian and 1700 military victims) as well as military troops. This unit gained ability to produce hundreds to thousands of kg's of BW-agents monthly. This work continued until the end of the World War II when the US government granted immunity to the Japanese Unit 731 leaders in exchange for the gained knowledge. The German researches followed the effects of vaccinations and drugs on prisoners infected with *Rickettsia prowazekii*, hepatitis A virus, or *Plasmodium* species. They showed interest in anthrax investigations. The USA started their BW program in 1942 by establishing the laboratory in Fort Detrick, Maryland, and various production facilities as well as dissemination and testing places were created around the country by the end of the war in 250 buildings with 6500 employees. In 1944 Churchill ordered from the US 500,000 anthrax-filled bombs with the intent to use them against the Germans. In that period were also produced cluster bombs with anthrax charge and the budget for this branch raised from 3.5 million to 60 million USD during the war [2].

After the WWII started the Cold War Period. Both super-powers as well as other countries continued aggressive development of BW programs. The agents and vectors of anthrax, tularemia, brucellosis, Q-fever, Venezuelan equine encephalitis, malaria, dengue, fungi that destroy useful plants (*Pirucularia orizae, Sclerotum rolfsii …*) were investigated in that period in USA as well as biological toxins and infective aerosols. In the 1950s started the first experiments concerning potential ethnic weapons. Offensive biological programs were officially halted in 1969 in the United States and Great Britain. Similar research and development on BW agents production and weaponization were conducted in the Soviet Union with special attention to viral agents. Under the"*BIOPREPARAT*" program were performed experiments with genetically modified microorganisms (resistance to antibiotics, environmental conditions, chimeras production like Ebolapox and other genetic modifications) [16]. Facilities included best known "*Vector*" in Koltsovo, then Sergiyev Posad, Kirov, Yekaterinburg and others. In 1954 was opened the testing base in Aral sea on Vozrozhdeniya Island where many agents were tested including anthrax and variola.

Scientists supported with state programs during the Cold War started playing the role of the God misusing the science progress especially in the field of genetic engineering and biotechnology. The world was faced with the danger of the first biological warfare. At that time, many BW attacks were made especially in the course of the Korean, Vietnam and the Afghanistan war as well as more than 500 accidents that killed thousands of people and hundreds of scientists occurred.

In parallel with the BW development, there was an increasing concern regarding the ineffectiveness of the Geneva Protocol so the new Convention on the Prohibition of the Development, Production and Stockpiling of Bacteriological (Biological) and Toxin Weapons and on their Destruction (BTWC) was signed in 1972 by the US, UK and Soviet governments, as well as by >100 other nations. BTWC entered into force in March 1975, and having been continuously reviewed [17]. It prohibits the development, production, storage, acquisition and transfer of biological agents and their toxins and the development of technologies for dispersal of biological agents for offensive military purposes ordering the destruction of existing stocks [18]. Today the BTWC is one of the three fundamental pillars of the international community's effort against WMD, along with the Nuclear Non-proliferation Treaty and the Chemical Weapons Convention.

Non-compliance or violation of this international treaty carries the heaviest legal and ethical conviction – a crime against humanity. The BTWC currently has 175 states-parties, although only several signatory states have not yet ratified the document. But the real ethical question is: are all of them honest in their implementation and intentions? The answer appeared immediately after the BTWC adoption and it was clear that the Convention existence did not prevent various states from developing BW research programs. Many accidents from that period have testified about the work with BW including the one in Sverdlovsk in 1979 with 68 reported deaths due to anthrax release from secret military facility or the death of researcher Nikolai Ustinov after laboratory infection with Marburg virus at "*Vector*" facility in Novosibirsk. The bacterial cultures from US ATCC basis were legally sold until 1989 to many countries including *Brucella* and *B. anthracis* sailing to Iraq. In the same time happened many terrorist actions with B agents using. The "*Order of the Rising Sun*" cult in 1972 obtained 30–40 kg of typhus to contaminate water sources of the US West Coast cities; Georgi Markov was killed in 1978 with ricin capsule placed in an umbrella; marxist group in West Germany planned the use of botulotoxin in 1980; the extreme group "*Dark Harvest Commandos*" used anthrax in 1981 to contaminate luggage of British politicians; the members of "*Rajneeshee*" group spread *S. typhimurium* bacteria over sallad bars in Oregon in 1984 to disrupt local elections; *Aum Shinrikyo* cult performed at least 9 terroristic attacks with B agents such as *Clostridium botulinum* and *B. anthracis* but with no proof of effectiveness; the anthrax attack in 2001 in USA resulted with 22 infected and 5 deads and subsequently open the questions concerning awareness, biosafety, biosecurity and other implications of bioterrorist actions) [2].

What is going on with ethics? The answer may partly provide time in which we live. Terrorism is today global evil that can be considered as a war crime and disease at the same time. And use of the WMD is real possibility as an old game under new rules and with new technologies. Biological weapons and agents cause the greatest concern among them since they are frighteningly fast being developed and spread and by their potential use can be achieved even strategic level effects. Science development and achievements provide new opportunities in knowing the physiological processes at the molecular level, diagnosis, therapy, the production of a vaccine all that particularly important for biodefence as well as new opportunities for development of BWs and bioterrorism. Thus, the results of scientific research, especially if it concerns fundamental natural processes, can have a wide range of desirable and undesirable technological applications. It is dual use dilemma, isn't it?

19.3 Genetic Engineering and Biotechnology in Development of BWs and Fight Against Them: Dual Use-Ethical Dilemma

In the last several decades, the world has witnessed a knowledge explosion in the life sciences based on an understanding of genes and how they work as well as the progress in biotechnology and nanotechnology. According to some estimation, in the life sciences, we now are where information technology was in the 1960s; more than any other science, it will revolutionize the twenty-first century. The understanding of the complex biochemical pathways that underlie life processes has the potential to enable a class of new biological agents engineered to attack distinct biochemical pathways and elicit specific effects, so the same science that may cure some of our worst diseases could be used to create the world's most frightening weapons [19].

Undoubtedly, scientific research has the potential to do good as well as harm in the same time and scientific knowledge that can benefit humanity can also be used to harm it especially through a wide range of desirable and undesirable technological applications. Sometimes it might be difficult, to decide at once which are desirable and which are not especially in fundamental sciences as it was with chain reaction and its use either for nuclear energy and nuclear bombs in the previous century [20].

Thus, let's remember that physics had the most important role in the development of WMD in the twentieth century, while the twenty-first century in this field will surely mark biology. The technology which raises the most concern today is the genetic construction and reconstruction of pathogenic organisms and viruses by biotechnological means. The great achievements of molecular biology and genetics have produced advances in agriculture and industrial processes and have revolutionized the medicine. The same results, however, pose a potential and unpredictable risk due to the possibility that these technologies could also be used to create the bioweapons of next generation [21] and can be considered as "dual use research of special concern". According to The US Science Advisory Board for Biosecurity this term refers to the research that provide knowledge, information, products, or technologies that could be directly misapplied to pose a significant threat with broad potential consequences to public health and safety, agricultural crops and other plants, animals, the environment, material, or security [22, 23]. Although all scientific fields appear to be affected by the dual-use dilemma, biological research seems to be more exposed to the risks of misuse, because it seems possible that in the future, technological advances will bring bioweapon development using synthetic techniques within the abilities of smaller and smaller groups without significant scientific expertise and serious state control.

Even if many area of biomedical research are affected by the dual-use scenario, unfortunately, bioethicists have to date had relatively little to say about security in general or the dual-use dilemma in particular. With a few recent exceptions, bioethicists neglect questions about whether it is ethical to produce and/or disseminate scientific knowledge [24]. This is ironic given the enormous amount of attention bioethics has placed on both:

- research ethics that is predominantly focused on the protection of human and animal research subjects,
- ethical, legal and social implications of genetics that are predominantly focused on potential environmental hazards of recombinant DNA research, genetic determinism, genetic testing, selective reproduction, genetic enhancement, cloning, stem cell research, DNA fingerprinting and the patenting of DNA sequences [25].

Thus, in order to protect humankind and prevent possible misuse of biotechnology and bioterrorism in the world it is necessary to establish an international consensus in bioethical approaches.

According to some authors the threat of synthetic biology misuse will be even greater than that posed by nuclear technology misuse because nuclear technology is likely to remain bulky and expensive, while the technologies required to produce bioweapons may become quite portable and cheap. The tradition of openness in the life sciences could be also a problem, meaning that much of the background knowledge relevant to synthetic biology is already in the public domain in contrast to advances in nuclear technology, which were often classified and confidential from the outset [26]. The following examples could be real confirmation of my words.

19.4 Experiments of Ethical Concern

In the article published in 2001 was described the accidental discovery of a group of Australian scientists who were attempting to produce an infectious contraceptive for mice, which periodically breed out of control in parts of Australia. The scientists spliced a single foreign gene encoding interleukin-4 (IL-4), molecule that regulates immune system into a mild mousepox virus in the hope of creating a genetically engineered sterility treatment. The effect was unexpected: creation of a so virulent mousepox strain that was able to kill both mice with natural resistance to mousepox and mice that had been vaccinated against the virus [27].

The study was carried out before the 9/11 attacks, when concern about the misuse of scientific outcomes was not very widespread. A disturbing implication of the experiment result is that adding an IL-4 gene might similarly increase the virulence of smallpox (or some other poxvirus that infects humans) and potentially allow the virus to overcome vaccination (which is our only defense against smallpox): that is, the same technique might also render viruses that do affect humans, such as smallpox that is global bio-threat again, vaccine resistant [28].

The biosafety level at the laboratory in Australia was sufficient to preclude accidental release of infected animals, but there is a real concern that genetically modified pathogens could be accidentally released without available countermeasures, such a release could pose a threat to endemic species and possibly to humans. This concern is likely to grow as biodefence research globally expands.

In the following study, scientists chemically synthesized a Polio genome in the absence of template by stringing together strands of DNA purchased over the Internet. Poliovirus is member of *Picornavirus* family. Its genome is a single-stranded RNA molecule of approximately 7500 nucleotides. The process was time consuming, but straight-forward. The cDNA was converted into RNA and then put not into cells, but into a mixture of proteins. The work, which was part of a program to develop biowarfare countermeasures, can be classified as the first creation of life in a test tube. The research resulted in the creation of a virus that paralyzed and killed mice developing a neurological disease both chemically and histologically indistinguishable from poliomyelitis [29]. Upon publication of results the researchers said *"the virus was made to send a warning that terrorists might be able to make biological weapons without obtaining a natural virus"*. Similar techniques might enable production of smallpox or Ebola [25]. The real dual-use issue for these experiments is the proof of principle that even more dangerous biological agents, could be made from scratch. The potential for misuse of DNA synthesis technology and genome sequence information is defined.

In the following study researchers used published DNA sequences to engineer a protein – known as SPICE –inhibitor of complement enzymes produced by the smallpox virus. The study revealed the ways in which, and the extent to which, this protein defeats the human immune system [30]. These findings may facilitate development of protective medicines based on the SPICE disabling, as a potent inhibitor of human complement system but they may also reveal ways to increase the virulence of the closely-related vaccinia virus (which is used in the smallpox vaccine) [25]. Dual use dilemma again?

The another case concerning reconstruction of the Spanish flu H1N1 influenza virus that caused the largest pandemic in human history in 1918/19 and killed 18–20 million people using preserved archived autopsy materials and a lung tissue of an influenza victim who had been buried and preserved in the permafrost of Alaska for RNA sequence generation and subsequent viral reconstruction as well as recombination with other viral strains using reverse genetics. The aim of this experiment was increasing understanding of biological properties responsible for the high virulence of the pandemic strain [31]. The experiment also indicated that the 1918 virus gene sequences were closely related to avian (H1N1) viruses more than any other H1N1 influenza strains. Although further research on the reconstructed virus may facilitate development of drugs and vaccines that provide protection against a major influenza pandemic, such a virus could also be used for nefarious purposes by malevolent actors [25]. This case also demonstrates how easy it has become to dispose of scientific results in order to create human pathogens and to achieve malevolent purposes.

The researchers from the J. Craig Venter Institute in Maryland, US, succeeded in synthesizing a full bacterial genome and in May 2010 announced the creation of the first living and replicating bacterium with a synthetic genome that differs only slightly from wild-type *Mycoplasma mycoides* [32]. It constitutes 'proof of principle' for the synthesis of a new bacterial chassis or other radically new bacteria which have not and could not have naturally existed. Intense media coverage followed, and the announcement ricocheted across the globe within hours as proponents and critics made striking claims about potential risks and benefits of this discovery and whether it amounted to an early-stage example of "creating life. Other synthetic biologists are seeking even more fundamental re-design of life and even developed two new bases which can be incorporated into DNA alongside the existing four bases, and can be replicated by naturally occurring enzymes. Though potential applications of the new bases remain unclear, the development suggests the possibility that we may be able to create DNA with highly unusual characteristics [26].

Synthetic biology has many potential applications and benefits in the fields of environment and energy production, health care (e.g. malaria drugs and gene therapy), and industry. But it also opens a number of issues associated with biosafety and laboratory biosecurity, the potential misuse of synthetic biology and many ethical, social and legal concerns about its impact on society, public health and the environment in addition to questions of the ownership, innovation, regulation and oversight [33].

In 1997, Dr. David Edwards described a revolutionary way to deliver aerosolized medicines using large porous carrier particles. The method dramatically increased the amount of inhaler-delivered drug that made it deep into the lungs [34]. While this finding attracted little media attention at the time, its dual-use implications became clear after five people died from inhalation anthrax in 2001 as the existing opportunity to "reverse engineer" an inhaled drug delivery system and increase terrorist ability to bypass natural defenses [35].

In the past, the only option for developing biological weapons was to select strains for certain qualities including environmental stability, lethality, ability to be aerosolized and antibiotic resistance. Now, with recombinant DNA technology and increased understanding of biological systems, desired characteristics can be engineered into pathogens and medical technologies applied to increase delivery of agents. So, the dual-use dilemma in biological science becomes an emerging issue in biodefense. Progress in biology and genetic manipulation will continue, and it is necessary to consider the existing biosecurity control measures and continue expand them to keep pace with technology development.

To the extent that important values are at stake, the dual-use dilemma is inherently ethical in nature. It is noteworthy, however, that most of the debates about the dual-use dilemma have primarily involved science and security experts rather than ethicists. At the other side, a huge number of journal articles and books on ethics and genetics had been written; but they include little, if any, discussion of the potential role of genetics in weapons-making.

19.5 Dual Use Dilemma As the Ethical Dilemma: Who Is Really Responsible?

The dual-use dilemma is undoubtedly an ethical dilemma since it is about promoting good in the context of the potential for also causing harm, e.g., for the killing of innocents. It is an ethical dilemma for the researcher because of the potential actions of others, e.g., malevolent nonresearchers who might steal dangerous biological agents, or make use of the original researcher's work. And it is a dilemma for governments concerned with the security of their citizens, as well as their health [28]. But is a scientist really responsible in this case? There is a basic conflict between the freedom of the researcher and the duty to prevent harm which is difficult to solve because it is unclear what the contribution to the relevant action is and to what degree this is controllable. As we are only responsible for something we can control, the point is to what degree the scientific community and the individual scientist can be considered to be able to control such effects [20]. Can scientists be responsible for the way the knowledge they produce gets used? Should we promote scientific research and dissemination of scientific knowledge that can be i.e. used to develop WMD? It is mainly the question for Governments/editors who have the power or authority to assist or restrict researchers' work. But which kind of responsibilities do scientific researchers have and which are the ethical principles of scientific research? Scientists have two kinds of specific responsibilities connected with their professional activities and relationships: those internal to science which require to respect the standards of practice approved by scientific community including responsibility towards the used animals and people involved in medical research and those toward society, which are referred to as scientists' social responsibilities. Because of all that mentioned, the ethics and responsibility of science should be an integral part of the scientists education and training in order to encourage young scientists to respect and adhere to the basic ethical principles and responsibilities of science [36].

The archetype of good scientific behavior is reflected in Merton's ethos of science suggesting that good scientific practice includes sharing of scientific results with others, because the science is "universal" so that knowledge claims are to be tested. The scientists should not only be involved in the production of new knowledge; they must be also committed to be critical towards knowledge claims raised by their colleagues, and are obliged to test their colleagues' results The scientific communities were also warned not to let their research projects be financed by power structures with special interests [37].

Recent advances in biotechnology have transformed the life sciences raising many often controversial issues. Breakthroughs can help humankind in many ways, but they invariably carry some risks, public concerns and, often, fears. Proponents of synthetic biology cite its potential to reduce our reliance on fossil fuels and transform medical care and human health, and other possible benefits. Critics express concerns about "playing God," threatening biodiversity, and threatening longstanding concepts of nature. With these unprecedented opportunities and achievements comes an obligation to consider carefully both the promise and potential perils that they could realize [38].

Research misconduct is defined as any behavior by a researcher, whether intentional or not, that fails to scrupulously respect high scientific and ethical standards and must have adequate consequences. Various types of research misconduct include fabrication or falsification of data, plagiarism, problematic data presentation or analysis, failure to obtain ethical approval by the Research Ethics Committee or to obtain the subject's informed consent, inappropriate claims of authorship, duplicate publication, and undisclosed conflict of interest [39].

19.6 Ethical Principles in Biomedical and Biodefence Research

Science is not only knowledge because what scientists do can have profound effects on humanity and on the environment. It is an important principle of morality that imposes us to consider the consequences of our actions for others, and scientists have the same responsibility.

Medical research is subject to ethical standards that promote and ensure respect for all human subjects and protect their life, health and rights – dignity, integrity, right to self-determination, privacy, and confidentiality of personal information of research subjects. Physicians must consider the ethical, legal and regulatory norms and standards for research involving human subjects in their own countries as well as applicable international norms and standards [40].

The ethical principles of animal research laid out the concept of the "Three R's": *replacement* of conscious living animals with nonsentient animals or materials, *reduction* of the number of animals used in an experiment or procedure, and *refinement* of the techniques used in order to decrease the incidence or amount of animal pain and distress. These concepts have been adopted by a number of scientists and many animal advocacy organizations and have been written into the laws of some countries [41].

A closer look to scientific practice leads to infer these four essential principles governing research and its duties towards the public. These principles of autonomy, beneficence, nonmaleficence and justice were argued to be mid-level principles mediating between high-level moral theory and low-level common morality, and they are very popular in writings about medical ethics [42].

Scientists must consider whether their work (or dissemination of scientific knowledge) carries more benefits than harms and whether it respects the principle of autonomy and justice. So, the idea of optimization is crucial because it emphasizes that the task is not to maximize the realization of any one value (such as protection against bioterrorism), or to achieve an acceptable trade-off between just two values (such as 'open science' and biosecurity), but rather to achieve an overall outcome that gives due weight to all relevant values [43].

A well established ethical principle within research ethics is the principle to prevent harm. An important distinction can be made between *intentional* and *unintentional harm.* The connection between intentional participation in developing and using biological weapons and harm is evident. This connection is less obvious in the case of legitimate and peaceful research subjected to unintended misuse and its potential to cause harm. Awareness is particularly important in this field. To be aware is an active process where you reflect upon your work and its potential risks and consequences. It also includes considering your role as a scientist in a national crisis or war situation and your relation to existing regulations [44].

In dual use dilemmas, scientists should consider whether the harm connected with their research is foreseeable, proportionally greater than the benefits and more easily achievable by other means and whether possible misuse is capable of being controlled, restricted and prevented by institutions or can lead to the development of WMD. The question here is not to which extent scientist are responsible for the intended effects of their work, but how far they are responsible for the foreseen effects and for their prevention and effort to predict them. It is not possible to eliminate risks of misuse but researchers should consider the best outcome as well as the worst in making decisions. The editors of scientific journals also have ethical responsibility. On some occasion an editor may conclude that the potential harm of publication outweighs the potential societal benefits and the paper should be modified, or not be published. So, journals and scientific societies can play an important role in encouraging investigators to communicate results of research in ways that maximize public benefits and minimize risks of misuse [45]. But voluntary self-regulation by scientists and editors about dual use dilemmas is unacceptable. Career advancement generally requires a strong publication record and scientist's interest may thus conflict with national security because scientists and science editors are not security experts. On the other hand, security experts are likely to be biased in favor of security values over scientific values [45]. There is also reason to doubt that governmental decision-makers will always have sufficient expertise to judge the scientific importance of publishing studies they might want to censor. This could be one more step down the path of liberty infringement in the name of "the war on terrorism" [25]. So, just as governmental officials are likely to have values biased in favor of security over the promotion of science, scientists and science editors are likely to be biased in favor of the promotion of science over security. Neither the scientific community nor the government has the competences to make final assessment regarding particular pieces of research and dissemination of information involving dangerous discoveries which have implications with WMD. The balance between competing interests necessarily requires that evaluations be handled by a group embodying experts in science and in (bio) security. Scientists might be best able to recognize a discovery's scientific or technical implications for the making of particular biological weapons, but they have no special expertise to determine the identity, abilities, or intentions of potential bioterrorists and to assess the possible security implications of biological attack [25].

Research undertaken for prevention or mitigation of biological threats can be used to cause harm by non-state terrorists or aggressive state actors or even by one's own government. This is another possible negative outcome of research, concerning governments and authorities [43]. Anyway there is a panoply of diverse ethical considerations that relate to biodefence including clinical testing of potential therapies and vaccines in young children and older adults, preventing unauthorized individuals from entering research laboratories that is also biosecurity issue, dual use dilemmas in publishing the results, developing of harmful technologies, allocation of educational resources etc. [5]. Taking all that into serious consideration, the US established the National Science Advisory Board for Biosecurity (NSABB) that has made steps to provide scientists with the information necessary for understanding security threats and make clear recommendations and criteria concerning the dual use research and publishing, considering that additional policy development in this area may be needed on the basis of previous experience [45–47].

So, content about dual use issues should be presented in a variety of contexts and scientific organizations. Education about dual use dilemmas should be incorporated into the channels through which life scientists already receive exposure to issues of responsible conduct, like biosafety, bioethics and research ethics, and responsible conduct of research education The adoption of a code of ethics for research could prevent the life sciences from becoming the dead sciences and ethics can be an important weapon to counter bioterrorism [48].

19.7 Ethical Issues in the Case of Eventual Bioterrorist Attack

In the case of eventual bioterrorist attack or biological warfare ethical principles and priorities could change. Some of the question with possible ethical consequences in that circumstances concerning allocation of resources and personnel-health care providers in specific circumstances and their personal approach in relation to personal health risks from maybe contagious infective diseases, risks for their families or fear for their own safety. Then there is a question of triage and treating patients either on first come, first treat basis or triage in order to save the greatest number of lives in disaster. To address these issues to the maximum benefit of potential victims, patients or exposed persons, we must first develop collective forethought and a broad-based consensus that these decisions must reach beyond. Critical decisions like these should not be made on an individual case-by-case basis. Physicians should never be placed in a position of individually deciding to deny treatment to patients without the guidance of a policy or protocol [49].

19.8 Conclusion

Many ethical questions are open in the field of biodefence, bioresearch public health, medical and environmental ethics as well as governmental and international relational ethics and even the ethics of the war conducting. International dialogue and consensus as well as ethical guidelines must be a part of strategy for the humankind survival in the world where bioterrorism is a real and global threat and biological weapons are dangerous enemies that do not recognize boundaries. Thus, we should listen to the words of famous scientist Thomas Edison who said *"Non-violence leads to the highest ethics, which is the goal of all evolution. Until we stop harming all other living beings, we are still savages."*

References

1. Block S (2001) The growing threat of biological weapons. Am Sci 89:28–38
2. Ristanovic E (2015) Bioterrorism:prevention and response. Odbrana Media Center, Belgrade
3. Samardzic S, Marinkovic T, Marinkovic D, Djuricic B, Ristanovic E, Simovic T, Lako B, Vukov B, Bozovic B, Gligic A (2008) Prevalence of antibodies to Rickettsiae in different regions of Serbia. Vector Borne Zoonotic Dis 8(2):219–224
4. Cekanac R, Mladenovic J, Ristanovic E, Lazic S (2010) Epidemiological characteristics of brucellosis in Serbia, 1980–2008. Croat Med J 51(4):337–344
5. Loike JD, Fischbach RL (2013) Ethical challenges in biodefense and bioterrorism. J Bioterror Biodef 12:2. https://doi.org/10.4172/2157-2526.S12-002
6. Szalados JE (2012) Triaging the fittest: practical wisdom versus logical calculus? Crit Care Med 40:697–698

7. Faden RR, Karron RA (2012) The obligation to prevent the next dual-use controversy. Science 335:802–804
8. Barras V, Greub G (2014) History of biological warfare and bioterrorism. Clin Microbiol Infect 20(6):497–502
9. Grmek MD (1979) Ruses de guerre biologiques dans l'Antiquité. Revue des Études grecques 92(436):141–163
10. Eneh OC (2012) Biological weapons – agents for life and environmental de struction. Res J Environ Toxicol 6:65–87
11. Wheelis M (2002) Biological warfare at the 1346 siege of Caffa. Emerg Infect Dis 8(9):971–975
12. Ristanovic E, Gligic A, Atanasievska S, Protic-Djokic V, Jovanović D, Radunović M (2016) Smallpox as actual biothreat: lessons learned from its outbreak in ex-Yugoslavia in 1972. Ann Ist Super Sanita 52(4):587–597
13. Robertson AG, Robertson LJ (1995) From asps to allegations: biological warfare in history. Mil Med 160:369–373
14. Christopher GW, Cieslak TJ, Pavlin JA, Eitzen EM (1999) Biological warfare: a historical perspective. In: Lederberg J (ed) Biological weapons. Limiting the threat. MIT Press, Cambridge, MA, pp 17–35
15. Barenblatt D (2006) A plague upon humanity: the secret genocide of axis Japan's germ warfare operation. Souvenir Press, London
16. Ristanovic E (2016) Health and security challenges of the 21st century: bioterrorism. ABC-časopis urgentne Med 16(1):8–19
17. Wheelis M, Dando M (eds) (2006) Deadly cultures: biological weapons since 1945. Harvard University Press, Cambridge, MA
18. Biological and Toxin Weapons Convention (BTWC). Available via http://www.opbw.org/convention/conv.html. Accessed 20 May 2017
19. Central Intelligence Agency Directorate of Intelligence (2003) The darker bioweapons future. Available via http://www.fas.org/irp/cia/product/bw1103.pdf. Accessed 21 May 2017
20. Ehni HJ (2008) Dual use and the ethical responsibility of scientists. Arch Immunol Ther Exp 56(3):147–152
21. Committee on Research Standards and Practices to Prevent the Destructive Application of Biotechnology (2004) Biotechnology research in an age of terrorism: confronting the "dual use" dilemma. National Academies Press, Washington, DC
22. Berger KM, Wolinetz C, McCarron K, You E, William So K (2012) Bridging science and security for biological research: a dialogue between Universities and the Federal Bureau of Investigation. Meeting report, Washington DC, February, 2012. American Association for the Advancement of Science, Washington, DC
23. Radakovic S, Marjanovic M, Surbatovic M, Vukcevic G, Jovasevic-Stojanovic M, Ristanovic E (2014) Biological pollutants in indoor air. Vojnosanit Pregl 71(12):1147–1150
24. Selgelid MJ (2010) Ethics engagement of the dual use dilemma: progress and potential. In: education and ethics in the life sciences: strengthening the prohibition of biological weapons. Available via http://press-files.anu.edu.au/downloads/press/p51221/pdf/ch012.pdf
25. Selgelid MJ (2009) Governance of dual-use research: an ethical dilemma. Bull World Health Organ 87:720–723
26. Douglas T, Savulescu J (2010) Synthetic biology and the ethics of knowledge. J Med Ethics 36(11):687–693
27. Jackson RJ, Ramsay AJ, Christensen CD, Beaton S, Hall DF, Ramshaw IA (2001) Expression of mouse interleukin-4 by a recombinant ectromelia virus suppresses cytolytic lymphocyte responses and overcomes genetic resistance to mousepox. J Virol 75(3):1205–1210
28. Miller S, Selgelid MJ (2007) Ethical and philosophical consideration of the dual-use dilemma in the biological sciences. Sci Eng Ethics 13(4):523–580
29. Cello J, Paul AV, Wimmer E (2002) Chemical synthesis of poliovirus cDNA: generation of infectious virus in the absence of natural template. Science 297(5583):1016–1018

30. Rosengard AM, Liu Y, Nie YZ, Jimenez R (2002) Variola virus immune evasion design: expression of a highly efficient inhibitor of human complement. Proc Natl Acad Sci 99(13):8808–8813
31. Keiser J (2005) Resurrected influenza virus yields secrets of deadly 1918 pandemic. Science 310(5745):28–29
32. Gibson DG, Glass JI, Lartigue C et al (2010) Creation of a bacterial cell controlled by a chemically synthesized genome. Science 329(5987):52–56
33. Chen Y, Yin Z, Shao Z, Xie Q (2015) The Defence of artificial life by synthetic biology from ethical and social aspects. J Coll Physicians Surg Pak 25(7):519–524
34. Edwards DA, Hanes J, Caponetti G et al (1997) Large porous particles for pulmonary drug delivery. Science 276(5320):1868–1872
35. Edwards DA (2002) Delivery of biological agents by aerosols. AICHE J 48(1):2–6
36. Cetto AM (ed) (2000) Science for the twenty-first century. A new commitment. Proceedings of the World Conference of Science, UNESCO, Paris, 2000. http://unesdoc.unesco.org/images/0012/001207/120706e.pdf
37. Hansen TB (2006) Academic and social responsibility of scientists. ISYP J Sci World Aff 2(2):71–92
38. United States Presidental Commission for the Study of Bioethical Issues (2010) New directions: the ethics of synthetic biology and emerging technology. Washington. Available via http://bioethics.gov/sites/default/files/PCSBI-Synthetic-Biology-Report-12.16.10.pdf. Accessed at 22 May 2017
39. Jain AK (2010) Ethical issues in scientific publication. Indian J Orthop 44(3):235–237
40. WMA (2013) World medical association declaration of Helsinki: ethical principles for medical research involving human subjects. JAMA 310(20):2191–2194
41. Bishop LJ, Nolen AL (2001) Animals in research and education: ethical issues. Kennedy Inst Ethics J 11(1):91–112
42. Beauchamp TL, Childress JF (2001) Principles of biomedical ethics, 5th edn. Oxford University Press, Oxford
43. Buchanan A, Kelley MC (2013) Biodefence and the production of knowledge: rethinking the problem. J Med Ethics 39:195–204
44. Kuhla F, Eriksson S, Evers K, Hoglund T (2008) Taking due care: moral obligations in dual use research. Bioethics 22(9):477–487
45. Selgedid MJ (2007) A tale of two studies: ethics, bioterrorism, and the censorship of science. Hast Cent Rep 37(3):35–43
46. National Science Advisory Board for Biosecurity (2007) Proposed frame work for the oversight of dual use life sciences research: strategies for minimizing the potential misuse of research information. Available at: http://oba.od.nih.gov/biosecurity/pdf/Framework%20for%20transmittal%200807_Sept07.pdf
47. van Aken J (2006) When risk outweighs benefit: dual-use research needs a scientific cally sound risk–benefit analysis and legally binding biosecurity measures. EMBO Rep 7(SI):S10–S13
48. National Research Council (2011) Challenges and opportunities for education about dual use issues in the life sciences. National Academies Press, Washington, DC
49. Pesik N, Keim ME, Iserson KV (2001) Terrorism and the ethics of emergency medical care. Ann Emerg Med 37(6):642–646

Chapter 20
Concluding Remarks

Vladan Radosavljevic, Ines Banjari, and Goran Belojevic

Abstract Terrorism is on the rise throughout the world and bioterrorism as the most sophisticated type of terrorism is consequently more probable now than ever. This book is a scientific and timely contribution to defence against bioterrorism. Its chapters cover all aspects of bioterrorism from the preventive point of view and give theoretical and pragmatic framework for a bioshield, which successfully encounters both the measures of biodefence and bioweaponing control. The hot topics of this book within primordial, primary, secondary and tertiary prevention of bioterrorism include: dual use research in synthetic biology, strategy of intelligence, emerging natural infectious disease, potential public health threat of refugee crises, methods of differentiation between a biological attack and other epidemics, plant derived antigens and antibodies, panic disorder during bioterrorist attack, safety and security regulations, bioreconnaissance and bioforensics, strategy of deterrence, environmental, nutritional and ethical aspects of bioterrorism.

After anthrax letter attacks in the USA in 2001, a robust public health system was widely proposed for successful coping with unusual epidemiological events, with enormous investments in microbiology, public health, clinical and pharmaceutical industry. This is possible only in the most developed countries in the world. But, are such tremendous investments really necessary? This book is looking for optimal answers and solutions in defence against bioterrorism.

The countries differ to a great extent in public health preparedness for a bioterrorist attack. Whether they are well prepared or still in the process of developing

V. Radosavljevic (✉)
Military Medical Headquarter, Ministry of Defence, Crnotravska 17, Belgrade, Serbia

University of Defence, Belgrade, Serbia
e-mail: vladanr4@gmail.com

I. Banjari
Faculty of Food Technology, Josip Juraj Strossmayer University of Osijek, Osijek, Croatia
e-mail: ibanjari@ptfos.hr

G. Belojevic
Institute of Hygiene and Medical Ecology, Faculty of Medicine, University of Belgrade, Belgrade, Serbia

V. Radosavljevic et al. (eds.), *Defence Against Bioterrorism*, NATO Science for Peace and Security Series A: Chemistry and Biology,
https://doi.org/10.1007/978-94-024-1263-5_20

biopreparedness policy, in almost all countries biopreparedness is a part of overall health and security emergency preparedness, with the ministries of interior and health being responsible. Involvement of military in national biopreparedness differs greatly from country to country.

The sooner an unusual epidemiological event (UEE) is recognized, the better. Effective surveillance mechanisms are therefore of utmost importance. Rapid access to reliable and accurate medical data is vital. Ideally, UEE surveillance systems should provide real-time data, generated by widely-available, rapid and cheap diagnostic tools. UEEs are challenges for prevention, detection, diagnostic and treatment capabilities of public health systems. It is also a challenge for logistic service and pharmaceutical industry, with more extensive decontamination requirements and an altered communication strategy.

From a legal point of view, a bioterrorism event differs to a great extent from a naturally occurring outbreak. Therefore, public health and law enforcement officials must plan and work together to coordinate their investigations and introduce new issues regarding ethics, human rights and obligations concerning bioterrorism.

In the era of synthetic biology and the possibility to create new artificial biological systems or to substantially change their genome and phenotype a dual use research dilemma arises – could its misuse anulate all beneficial effects for mankind and how realistic is the possibility of a scientist-bioterrorist link. This is one of the challenges of the strategy of intelligence. Improvements in national public health systems for efficient coping with emerging naturally occurring infectious diseases and the epidemiological challenges of migration crisis are of crucial importance for defence against bioterrorism. The higher the people's trust in their national public health system the lower the possibility of a massive panic disorder as a consequence of a bioterrorist attack – one of the main aims of bioterrorists. Through safety and security regulations, up-to-date bioreconnaisance of environmental and food chains, bioforensics, digital surveillance systems and methods for a fast differentiation between biological attacks and other epidemics a mighty barrier against bioterrorism is being built.

This book provides a framework for a bioshield against bioterrorism. This bioshield is **available**, because it is mainly based on existing public health and security capacities. By improving their public health systems all countries may contribute to global biosecurity. This bioshield might be efficient due to its comprehensiveness. With four levels of prevention, as well as strategies of bioweaponing control and deterrence of bioterrorism, this bioshield is up-to-date and ready for implementing. It contains doctrinarian, strategic, operational and tactical instructions against bioterrorism and any other UEEs. Finally, this bioshield is easy for understanding and consequently pragmatic for decision making. Such system may significantly diminish the consequences of naturally occurring outbreaks, bioterrorist attack or other UEEs and may also contribute to deterring adversaries from pursuing attacks.

Printed by Printforce, the Netherlands